The Large Had

Martin Beech

The Large Hadron Collider

Unraveling the Mysteries of the Universe

 Springer

Martin Beech
McDougall Crescent 149
S4S 0A2 Regina Saskatchewan
Canada
martin.beech@uregina.ca

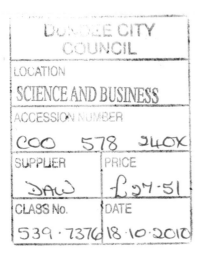
ISBN 978-1-4419-5667-5   e-ISBN 978-1-4419-5668-2
DOI 10.1007/978-1-4419-5668-2
Springer New York Dordrecht Heidelberg London

Library of Congress Control Number: 2010931433

Springer is part of Springer Science+Business Media (www.springer.com)

*This book is dedicated to my parents,
Leonard and Irene Beech.*

*For their many years of patient support
and encouragement, I am grateful.*

Preface

The Discarded Image was the last book that C. S. Lewis, perhaps better known for his *Narnia* series of stories, wrote. In this final tome, published in 1964, Lewis reflected upon many decades of lecturing, scholarly research, and philosophical thought.

The "image" that Lewis was concerned with was that of the medieval universe, and specifically its complete, compact, and fully determined form. Indeed, the image of the medieval universe is the very antithesis of the one that we have today. Although our universe is inconceivably large, nowhere near fully surveyed, only partially explained, and full of surprises, it does have one parallel with the medieval image: all is connected, and as every medieval astronomer knew, within the microcosm is a reflection of the macrocosm and *visa versa*.

The Large Hadron Collider (LHC) experiment now under commission at CERN [Conseil Européen pour la Recherche Nucléair][1] is just a modern-day continuation of this basic ancient tenet, and even though conceived and constructed to test the best present theories of particle physics, the results from the LHC will provide fundamentally new insights into the origin of the universe and its observed large-scale structure. All is connected.

This book will be concerned with the fleshing out of a new image that binds together the macrocosm (the universe) and the microcosm (the world of elementary particles). Our task in the pages that follow will be to "un-weave" the fabric of the universe, and to thereby tease out the intricate strands that connect the Standard Model of particle physics (and its many present possible extensions) to the observed cosmos around us. For indeed, it is now abundantly clear that once in the deep past, some 14 billion years ago, the vast expanse of the observable universe (with a present diameter of order 93 billion light years) was minutely small – a compact cloud of raw bubbling energy, full of future potential, arguably quixotic, and evolving on a mercurial timescale faster than the blink of a tomcat's eye. Out of the

---

[1] This was the provisional title for the organization at its founding in 1952. In 1954, however, a name change to *Organisation Européenne pour la Recherche Nucléaire* – European Organization for Nuclear Research – was agreed, but the original acronym, for reasons that are not clear, was kept.

primordial microcosm grew the macrocosm that is the universe of today, with its associated flotsam of galaxies, stars, planets, and life.

Verily, since it seems only right to use exulted tones, science and the work of countless observers and cloistered theoreticians have brought us quite an image to deal with. The modern-day image of the universe is full of incredible and unknown wonders. It is an image that would have thoroughly appalled the medieval scholar, not least because of its vast emptiness but also because its dominant components are entirely unknown to us. The mystery of dark matter (most definitely detected, as we shall see in Chapter 5, yet entirely unseen) and the possible existence of dark energy (discussed in Chapter 6) represent the most profound scientific problems of our age. And although it is not known what physical effects underlie these phenomena (yet), the LHC, by smashing head-on the nuclei of two lead atoms, will transport our understanding back to those moments that prevailed just 10^{-25} s after the Big Bang occurred – back to a time before stable matter even existed. Indeed, The LHC, in addition to its many other incredible properties, is also, in essence, a time machine, and this massive collider will enable researchers to study, for the merest fraction of a second, the primordial fire (the so-called quark–gluon plasma) out of which everything that we can now see and feel initially appeared.

The LHC experiment will not only provide researchers with insights as to why the universe has the characteristic form that it does (being made of something rather than absolutely nothing), it will also look for signs of the much anticipated Higgs boson, one of the key theoretical components of the Standard Model of particle physics, since it is generally believed that it is through interactions with the Higgs field that the various elementary particles acquire their mass; and this is no mere ivory-towered problem – without the Higgs (or some similar such mass-generating process) there would be no matter and no us.

Although the history and origins of the LHC will be described in greater detail in Chapter 2, we should provide at least a few words about its incredible properties before moving into our discussion on the basic properties of matter. The LHC is a machine – perhaps symbiotic complex is a better term for it – that can only be described in superlatives. As the medieval astronomer would have marveled at the great cathedrals of Paris and Rome, so the LHC is the pinnacle of modern experimental physics writ large on the landscape (actually under the landscape, as we shall see). We can do little but wonder at the LHC; its intricate yet paradoxically parsimonious structure, along with its sheer scale, leaves us humbled.

Indeed, the mind reels at the very thought that such machines can even be constructed. The numbers speak for themselves: the main accelerator ring is 26,659 m in circumference; the particle beams are manipulated by 9,593 super-cooled magnets that operate at a temperature of –271.3°C (just 1.9° warmer than the coldest temperature that anything can possibly be in the universe); and the system contains about 7,600 km of electrical cable, using strands of wire that if unraveled and joined end to end would stretch to the Sun and back five times over. When fully operational the LHC experiments will generate about 700 MB of data per second, or some 15 petabytes of data per year – enough digital data, in fact, to produce a 20-km-high stack of CDs every year.

The scale is grand, the structure is colossal, the task is Herculean, but the results from the LHC experiment could confirm and also re-write particle physics as it is presently known. The past, present, and future story of the LHC is and will be a fascinating one to follow, and it is an unfolding epic that could conceivably be destined to change our understanding of both atomic structure and the cosmos. All is connected, and within the macrocosm resides the microcosm (Fig. 1).

Fig. 1 An aerial view of the CERN complex. Set amidst the verdant fields of the French-Swiss border, the loop of the LHC collider ring is shown in the image center and foreground, with Lake Geneva and Mont Blanc, the tallest mountain in Europe, in the far distance. (Image courtesy of CERN)

The expectations of the medieval scholar were entirely different from those of today's scientist. Although our forebears would have held out zero expectation of discovering novelty within the universe (and within the properties of matter, for that matter), the modern observer fully expects to find new celestial objects and unexpected behaviors. The LHC is the tool that likely will reveal the new and the novel, and we can certainly expect that not only will our appreciation of the universe be very different a decade from now, but so, too, will our understanding of fundamental particle physics. It is almost certainly going to be a wonder-filled journey.

However, this journey has its associated risks. The LHC has been designed to explore the unknown, and some scientists have suggested that the CERN researchers may be on the verge of opening a veritable Pandora's Box of trouble. Literally, it has been argued that the LHC could release a host of exotic "demon" particles from beyond the borders of known physics – miniature black holes and so-called strangelets that some believe could pour from the LHC and potentially destroy Earth. These are frightening claims, and they must be considered carefully. How safe is the LHC, and how can we be sure that is doesn't pose a serious threat? These are questions that must be answered as we move forward, not only in the following pages of this book but also as we move into the future exploring the microcosm at ever higher energies.

Contents

1   The Story of Matter   1
  A Few Searching Questions   1
  The Smallest of Things   5
  Mysterium Cosmographicum   9
  A Particle Primer 11
    Thomson's Plum Pudding and an Unexpected Rebound 12
    The Quantum World and the Bohr Atom 15
  The New Quantum Mechanics 21
    Exclusion 23
    Fermi's Little Neutron 24
  Three Quarks for Muster Mark! 29
  Building the Universe 31
    The Matter Alphabet 31
    We Are of the Stars 32
  The Hubble Deep Field 38
  Moving Forwards 40

2   The World's Most Complicated Machine 41
  The End of the Beginning 42
  Disappointment and Setback 46
  Court Case Number 1:2008cv00136 48
  Afterwards 50
  Overview: A Proton's Journey 51
  The Journey to the LHC 56
  Collider Basics 57
  The Detectors 62

3   The Standard Model, the Higgs, and Beyond 71
  Generation the First – An Acrostic 71
  Feeling the Force 74
  The Higgs Field – Achieving Mass 78
  Feynman Diagrams 79
  Searching for the Higgs 82

Supersymmetry   86
Exotica: Going Up, Going Down   88

4 The Big Bang and the First 380,000 Years   91
The Big Bang   95
The Critical Density and Ω   98
The Microwave Background   99
Primordial Nucleosynthesis 103
Inflation, Flatness, Horizons, and a Free Lunch 105
The Quark–Gluon Plasma 110
ALICE: In Experimental Wonderland 113
Matter/Antimatter: It Matters! 115
Getting to the Bottom of Things 117

5 Dark Matters 121
Interstellar Matters 122
Where Are We? 126
Unraveling the Nebula Mystery 128
The Galaxy Zoo 130
The Local Group 132
Galaxy Clusters 133
Where's the Missing Mass? 135
All in a Spin: Dark Matter Found 137
Gravitational Lenses and Anamorphic Galaxies 142
Some Dark Matter Candidates 147
    The Neutralino 147
    Looking for MACHOs 149
    DAMA Finds WIMPS, Maybe 149
    CDMS Sees Two, Well, Maybe 151
    Bubbles at COUPP 151
    CHAMPs and SIMPs 153
    PAMELA Finds an Excess 155
    Fermi's Needle in a Haystack 156
ADMX 157
Euclid's Dark Map 157
The MOND Alternative 159
Dark Stars and Y(4140) 160

6 Dark Energy and an Accelerating Universe 163
The Measure of the Stars 165
An Expanding Universe 167
Death Throes and Distance 169
Future Sun – Take One 170
The Degenerate World of White Dwarfs 172
Future Sun – Take Two 175
The Case of IK Pegasus B 177

High-Z Supernova Surveys   179
Dark Energy and ΛCDM Cosmology   180
A Distant Darkness   184
Testing Copernicus   186

7   The Waiting Game   189
Hoping for the Unexpected   189
Massive Star Evolution   190
The Strange Case of RXJ1856.5-3754 and Pulsar 3C58   198
Small, Dark, and Many Dimensioned   203
This Magnet Has Only One Pole!   211
These Rays Are Truly Cosmic   214
Looking Forward to LHCf   218
The King Is Dead! Long Live the King!   219

Appendix A
Units and Constants   223

Appendix B
Acronym List   225

Appendix C
Glossary of Technical Terms   229

Index   233

About the Author

Dr. Martin Beech is a full professor of astronomy at Campion College at The University of Regina in Saskatchewan, Canada. He has published many scientific research papers on stellar structure and evolution and several books on astronomy. Asteroid 12343 has been named in recognition of his research on meteors and meteorites. This is Beech's third book for Springer. He has already published *Rejuvenating the Sun and Avoiding Other Global Catastrophes* (2008) and *Terraforming: The Creating of Habitable Worlds* (2009).

Chapter 1
The Story of Matter

A Few Searching Questions

Science is all about asking questions and looking for logically consistent explanations of what is observed. But, more than simply the searching for and finding of answers, the quest of the scientist is never over. There is always another question that can be posed, and there is always a different way, perhaps a better way, of explaining an observation.

This ceaseless process of searching, testing, pushing, and pulling at an idea and questioning is exactly what makes science so powerful, and it is also what makes it so successful at explaining the world that we see around us. For the scientific approach is by far humanity's best choice if a meaningful understanding of the universe and how it works is ever to be achieved. All other approaches lead either to fantasy or dogma.

Certainly, wrong explanations have been, and no doubt are still being, proposed by scientists, and on occasion entirely wrong ideas have been accepted for long periods of time as realistic explanations to some phenomena. But eventually, inevitably, the scientific process is self correcting. Science is ultimately ruthless, totally impartial, and completely devoid of feeling, but for all this it allows us to creep forward, inching ever closer towards a finer and better understanding of the marvelous universe and the many wonders that reside within it.

Given science is concerned with asking questions, then let us search the depths of our human senses and ask what the limits of our perceptions are. What, for example, is the smallest thing that you can see with your unaided eye? Certainly a period (.) is visible, but what about something half its size? Perhaps this is still visible to some readers; the author's aged and far less than perfect eyesight, however, would struggle to pick out such a miniscule ink-speck. Certainly something, say, one-tenth the size of a period would be below the ability of the best human eye to detect with any certainty. For the unaided human eye, therefore, the limit of smallness is achieved at about 0.05 mm.

From the inanimate viewpoint of the electron located in the ground state of the hydrogen atom (the meaning of all this will be explained later), the naked-eye limit of human perception corresponds to a distance that is about 500,000 times larger

M. Beech, *The Large Hadron Collider*, DOI 10.1007/978-1-4419-5668-2_1,
© Springer Science+Business Media, LLC 2010

than the orbit it occupies. The atom and the many exotic subatomic structures that will be described in this book are all entirely invisible to our naked eye, and yet science unequivocally tells us that they exist, and that they are real entities with measurable and understandable properties. Here, indeed, lies the power of scientific inquiry, since it can take us far beyond our body's ability to sense, and it affords us a deep and searching pair of artificial eyes with which to observe new and fantastic domains.

What is the most distant object that can be seen with the unaided eye? In the prairies of Canada, where this author lives, the distant horizon is about 33 km away when viewed from the top of the 25-story, 84-m-high Regina Delta Hotel. From the top of Chomolungma (Mt. Everest), the highest mountain in the world, the mountain climber's distant horizon (ignoring clouds and other mountains in the way) might stretch to 330 km.

We should not be so parochial in our views, however, as the painter John Ruskin reminds us that "mountains are the beginning and end of all natural scenery." The Moon, the Sun, and the planets out to distant Uranus are all visible to the unaided eye. Although discovered fortuitously by William Herschel and first recognized as a planet in 1782, Uranus is just visible to the unaided eye if one knows where to look. At closest approach to Earth, Uranus is about 18 astronomical units (AU) away, a distance equivalent to about 2.7 billion kilometers, or over 8 million times further away from us than the most distant horizon (that from Mt. Everest) visible on Earth. The nearest star, next to the Sun, of course, that is visible to the unaided eye is Alpha Centauri (which is actually a binary star; see Fig. 1.1), and it is 1.35 parsecs away, or about 41.5 million million kilometers. We are now 126 billion Everest horizons away from Earth.

We can search for stars fainter than Alpha Centauri in the sky, and these will mostly be further away, but the ultimate span over which our unaided eye can see is to a distance beyond any star in our Milky Way Galaxy and even beyond the stars in the nearest dwarf galaxies. The depth of human eye perception stretches all of the way to the spiral-shaped Andromeda Galaxy. Located in the constellation of Andromeda (Fig. 1.2), what looks like a faint fuzzy patch of light to our eye is about 736 kiloparsecs distant. Andromeda probably affords us a doppelganger image of our Milky Way Galaxy, and by viewing it (Fig. 1.3) we get a proxy grandstand view of our galactic home, a home that we will have much more to say about in Chapter 5.

The distance estimate for the Andromeda Galaxy completely escapes our sense of scale. It is 23 billion billion kilometers away, or equivalent to 69 million billion Mt. Everest horizons. It is remote on a scale that leaves us almost breathless and reeling. It takes light about 2.4 million years to traverse the distance between Andromeda and Earth, and the light that we see from the Andromeda Galaxy today started on its journey long before we, as *homo erectus*, had even evolved.

The world of human perception is sandwiched between the limits set by the tiny period and the Andromeda Galaxy, limits corresponding to about 5×10^{-5} m on the small side and 2.3×10^{22} m on the long. But the limits of our human senses are dwarfed by the atomic and universal scales that science allows us to explore;

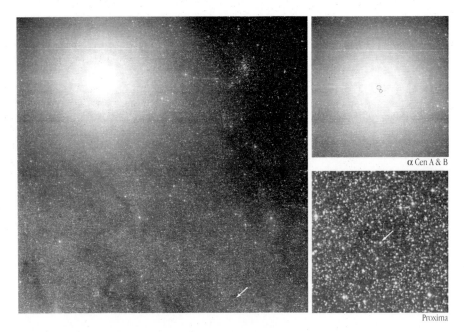

α Cen A & B

Proxima

Fig. 1.1 Alpha Centauri, the nearest star, after the Sun, visible to the unaided eye. The "star" is part of a binary system in which two stars orbit each other at a distance of about 24 Astronomical Unites (AU), with a period of about 80 years. The star Proxima Centauri is actually closer to us than Alpha Centauri, but it is not visible to the unaided eye. The images shown here were obtained with the European Southern Observatory's 1-m Schmidt Telescope. (Image courtesy of ESO)

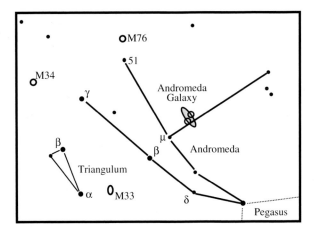

Fig. 1.2 A star map indicating the sky location of the Andromeda Galaxy, the most distant object visible to the unaided human eye

indeed, as Blaise Pascal was to reflect with both wonderment and vertiginous fear, we occupy that region that teeters between the two abysses of the infinitely large and the infinitely small.

The smallest size that physicists believe to be meaningful, by which it is meant that known physical theories should apply, is that of the Planck length. Named in honor of German physicist Max Planck, one of the pioneering founders of quantum mechanics in the early twentieth century, the Planck length corresponds to a minuscule distance of some 1.6×10^{-35} m.

The Planck length is a rather curious number, composed as it is from a combination of the universal gravitational constant G, Planck's constant h, and the speed of light c. For all this apparently abstract construction, however, the Planck length is the fundamental scale below which known physics no longer applies, and it is the new realm of the presently unrealized, but greatly sought after, domain of quantum gravity.

Between the domains of the world observable to humans and the world over which known physics applies we encounter a mind-numbing scale of magnitudes. The realm of known physics operates over scales 30 orders of magnitude smaller than our eyes can see. The observable universe also swamps our human eye limit to Andromeda by a factor of about 40,000. Remarkably, on the scale of our human senses, the domain of the inner, atomic realm outstrips that of the greater universal one.

The Smallest of Things

The philosophers of ancient Greece began the long struggle that has become modern science. The struggle, then and now, was not one relating to existence, but was rather the struggle to tease from nature a rational understanding of its properties. Their ideas on how the universe was ordered and how it came into being were no less remarkable than our own, but the modern-day concepts are more rigorously based upon observations and experimental measurements – observations and experiments, of course, that the ancient philosophers had no way of making.

Similar to the situation that prevails today, the ancient Greek philosophers did not want for theoretical ideas about the origin of the heavens; the problem then, as now, was rather one of differentiating between rival possibilities and agreeing upon an underlying physical theory of how nature works. By circa 400 B.C. it was generally agreed, however, that a spherical Earth was located at the center of a spherical cosmos, and that there were two types of celestial motion, that of the stars and that of the planets. In each case the motion was taken to be circular, and eternal, but the rate of motion was different in each case. Likewise, by about the same time it was generally agreed that objects in nature were composed of atoms, or elemental building blocks that were extremely small and that could not be further subdivided. One of the greatest (but perhaps more difficult to decipher) accounts of nature and the origins of cosmic order is that given by Plato in his speculative dialog *Timaeus* written circa 300 B.C.

The key to understanding nature, according to Plato, was one of identifying the good or benefit in its arrangement. This rationale stemmed from Plato's philosophical dictate that the universe had been brought into existence by a benevolent demiurge who had strived throughout to make its construction as near perfect as possible. According to Plato, therefore, things happen in the universe because the various objects within it are seeking to find their best possible configuration. Plato's cosmology is accordingly teleological and yet fully consistent with Socrates's (Plato's great teacher and mentor) mandate that a cosmological model should seek to explain why the cosmos is so structured and why its contents are so arranged in the best possible way.

To Plato it was in the nature of solid, earthy matter, for example, to fall or move towards the center of the universe, and by this reasoning he explained why Earth was spherical, since this corresponded to the tightest, most even packing of all the earthly material, and it also explains why Earth was centrally located within the cosmos – the reason being that Earth is made up of earthy matter. In this latter sense, for Plato, Earth is located at the center of the universe not because it is special, but because of what it is made of. Plato also considered the universe to be alive, an idea echoed, in some sense, in more recent times in the writings of James Lovelock and the Gaia hypothesis in which Earth is considered to be a large-scale, self-regulating, living organism. Here, in fact, is a nice example of an old, discarded idea being re-invigorated in the modern era.

Although we no longer give credence to Plato's living universe (because of our better understanding of what the expression "being alive" actually means and entails), Lovelock's Gaia hypothesis is much more rigorously defined and also much more restricted in its scope than Plato's world as described in the *Timaeus*. Not only do we continue to find the microcosm reflected in the macrocosm in the modern era, we also continue to find old ideas reflected and re-invented in the new.

Plato's living cosmos was infused with what he called the world-soul, which can be thought of as an animating force and intelligence that guides change to work towards the better good. Plato's spherical universe was essentially divided into two realms, that of the heavens and that of Earth. Above the spherical Earth's upper fiery-air region (the atmosphere to us) resided the perfect realm of the planets and the celestial sphere. The planets, which to the ancient Greeks constituted the Moon, Venus, Mercury, the Sun, Mars, Jupiter, and Saturn, were deemed to move along per-fectly circular paths around the center of the universe and from Earth were observed to move within the zodiacal band of constellations wrapped around the celestial sphere. The celestial sphere had its own perpetual motion, and it was this primary motion that caused the stars, considered by Plato to be living entities that were divine and eternal, to move around a stationary Earth.

In the Earthly, sub-lunar realm conditions were much less pristine than those encountered in the greater cosmos and certainly not eternal. With respect to matter, Plato argued in the same vain of Empedocles, who lived circa 450 B.C., positing the existence of four basic elements: earth, air, fire, and water. All matter in the sub-lunar region was made up of combinations of these basic elements. Objects in the celestial realm were composed of a special pure and incorruptible sub-stance called quintessence. In turn the basic elements were composed of minute particles (atoms), each of which had a special three-dimensional form. Indeed, Plato described the elemental atoms in terms of the regular or Platonic solids (Fig. 1.4).

The Platonic solids are special in that they are the only solids (or more correctly polyhedra) that can be made with the same generating shape for all of their faces. The hexahedron (or cube), for example, is made up of six squares, while the octa-hedron is made up of eight equilateral triangles (see Table 1.1). What Plato knew and presumably liked about these polyhedra, apart from their visual appeal, was that only five of them can possibly exist; there are no other regular polyhedra com-posed of more complex face panels. By associating the elemental atoms with the regular polyhedra, therefore, Plato was assured of there being a finite generating set of atoms. In this manner, each of Plato's six atomic polyhedra had an associated elemental composition (Table 1.1, last column). The element of earth, for exam-ple, corresponded to the cube, while that of water corresponded to the 20-sided icosahedrons, and so on.

As we shall see towards the end of this chapter Plato's list of atoms (just 5) is very small compared to our modern-day list. His list, however, is nonetheless a matter

Fig. 1.4 The platonic (also regular) solids. These are the only five polyhedra that can be made entirely of similar-shaped polygonal faces

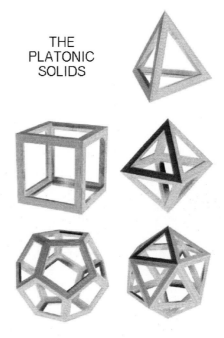

THE
PLATONIC
SOLIDS

Table 1.1 Characteristic properties of the Platonic solids. The first column gives the name for each of the polyhedra. The second identifies the generating polygon, and the third shows how many faces each polyhedron has. The last column is the element association given by Plato

Polyhedron	Generating Shape	Faces	Association
Tetrahedron	△	4	Fire
Hexahedron	☐	6	Earth
Octahedron	△	8	Air
Dodecahedron	⬠	12	Quintessence
Icosahedron	△	20	Water

alphabet that describes how all possible substances can be put together, and it also enables a basic alchemy (chemistry is far too grand a word for it) to be described.

Transformation of one element into another was possible, Plato argued, provided the generating faces of the various participating polyhedra were the same. One could take two atoms of fire, for example, and break them down into 8 equilateral triangles (each tetrahedron having 4 triangular faces), and then reassemble them as an octahedron, thus creating an air atom. We could write this reaction as 2Fire ⇒ 1Air. Many other transformation are possible, with, for example, 3Fire + 1Air ⇒ 1Water, or 2Water ⇒ 5Air.

Now, although Plato, as far as we know, didn't perform any experiments to see if such transformations could really come about, his basic outlook was not greatly dissimilar to that of today's chemist. Indeed, the science of chemistry is concerned with understanding the relationships and interactions between molecules, which are made up of atoms, with the atoms being from the Periodic Table of Elements. Once one type of molecule has been produced, then under specific conditions it can interact with a second type of molecule to generate a third type of molecule, and so on. In this basic manner all of the various solids, liquids, and gases can be built up and explained as being vast collections of specific atoms, with the different kinds of atoms being derived from a finite list.

In Plato's transformation theory, just as in the case of modern chemistry, some reactions are not allowed. A glance at Table 1.1 indicates the problem. Only the elements fire, air, and water could undergo transformations because the generating faces of their associated polyhedra were triangles. The earthly elements, in contrast, could not be transformed directly into fire, air, or water since they were composed of hexaheda that were generated by squares. Likewise quintessence, associated with the 12 pentagram-faced dodecahedron, cannot be transformed into any other form of matter. Within Plato's alchemical theory of atoms and elements, therefore, there are some forms that are stable, apparently forever, while others are more mutable and can switch form from one to another.

The idea of the basic elements was further expanded upon, especially in the medieval era, to include additional qualities such as being hot, dry, cold, or moist. These additional attributes resulted in the development of a diagnostic medicine, with the human body being brought into the cosmic fold. Indeed, the human body was deemed to be under the influence of four humors: *cholericus* (hot + fire), *melancholicus* (dry + earth), *phlegmaticus* (cold + water), and *sangineus* (moist + air). In a healthy body these four humors would be in balance, but in an unhealthy body one or more of the humors were held to be out of balance, and an appropriate medicinal step was required to restore both equilibrium and health. Bloodletting, for example, might follow a diagnosis of an excess of the sanguine humor – the moist + air combination that resulted in the formation of blood. Since, however, the flow of blood was deemed to be influenced by the lunar phase, a doctor might be reluctant to let blood in a specific region of the body at the time of a full Moon, and this accordingly introduced a role for the various planets. (The Moon, you'll recall, was considered a planet at that time.)

Indeed, each part of the body was associated with one of the 12 zodiacal constellations; the heart, for example, was ruled over by the constellation of Leo, while the knees were ruled over by Capricorn. With these associations having being adopted, an illness could be treated according to which planets were in which specific constellation. It is for this very reason that one of the most important courses that a medieval doctor would take during his university training was astrology. Remarkably, the four elements and four humors model brought the very universe into the workings of the human body, and, once again, we find the idea that within the microcosm is the macrocosm.

Clearly much has changed in our understanding since the time of the ancient Greek philosophers, but some of their essential outlook is very familiar to us in the modern era. The idea that matter is composed of extremely large numbers of very small basic building block, or atoms, that can bond together and turn into other forms under certain circumstances is exactly what we call chemistry. Our modern rules for transformation are more clearly defined and understood, but the basic idea is the same. This constancy of an underlying idea (with numerous modifications, admittedly) over many hundreds, even thousands, of years of human history is very rare, and the same cannot be said for our understanding of the cosmos.

Mysterium Cosmographicum

The *Mystery of the Heavens* was Johannes Kepler's first book, and it was written with a fearless passion and the energetic enthusiasm of youth. While historically this text, published in 1596, is less well known than his other great works relating to the refinement of Copernicus's heliocentric cosmology (to be discussed in Chapter 4), it is a wonderful book crammed full of mathematical insight and speculation.

Remarkably, we know exactly when Kepler had the seed idea that resulted in the new cosmological model presented in the *Mysterium*. The flash of insight occurred on July 19, 1595, during an astronomy class in which Kepler was talking about triangles and the properties of their inscribed and circumscribed circles (Fig. 1.5). The example that Kepler was considering during that fateful July class concerned the motion of Jupiter and Saturn around the zodiac, and being the great mathematician that he was he noticed that the ratio of the orbit radii for Saturn and Jupiter was the same (well, nearly so) as that corresponding to the radii of the circumscribed and inscribed circles of an equilateral triangle. This observation relating to the spacing of the obits of Saturn and Jupiter set his mind reeling, and he reasoned that perhaps the other planets are spaced according to the circumscribed circles that can be

Fig. 1.5 The circumscribed and inscribed circles to an equilateral triangle. The ratio of the radii of these two circles is almost identical to the ratio of the orbits of Saturn and Jupiter

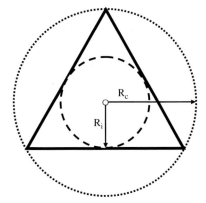

constructed around other plane figures such as the square or pentagon. He enthusiastically worked through the calculations but soon came up short. The ratios just didn't work out as an explanation for the observed spacing of the planets. Undaunted, however, and giving clear testament to his skill as a mathematician, Kepler was soon able to show that the orbital spacing of the planets could very nearly be explained according to the nesting, not of plane figures such as the square and triangle but according to the three-dimensional spheres that can be inscribed and circumscribed around the Platonic solids (Fig. 1.6).

The idea was beautiful, and the harmony exceptionally pleasing to Kepler, and in spite of his other great contributions to astronomy he never quite gave up on this early idea; it was his *idee fixe*. How could something so mathematically delightful, he reasoned, not be the true model upon which the universe (the planetary system as we would now call it) have been constructed by the great and omnipotent maker?

The problem with Kepler's cosmology, however, is that it just doesn't quite agree with the observed spacing of the actual planetary orbits, and as Kepler was to reveal a few years later in his life, it did not account for the fact that planetary orbits are elliptical and not circular (Table 1.2). One point that Kepler felt was particularly elegant and compelling about his new cosmological model, however, was that it offered a clear explanation as to why there were only six planets (the planets Mercury through to Saturn; the next planet outwards, Uranus, wasn't to be discovered until 146 years after Kepler's death). Given that there are five Platonic solids,

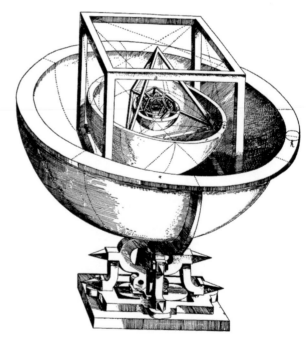

Fig. 1.6 Kepler's cosmological model designed according to the nesting of the Platonic solids

Table 1.2 Kepler's cosmological model based upon the nesting of spheres set between the Platonic solids. The fourth column shows the ratio of the radii corresponding to the circumscribed and inscribed spheres generated by the various polyhedra (third column). The last column shows the actual ratio of orbital radii

N	Planet	Circumscribed Polyhedron	R_C/R_I	R_N/R_{N-1}
1	Mercury	–	–	–
2	Venus	Octahedron	1.73	1.868
3	Earth	Icosahedron	1.26	1.383
4	Mars	Dodecahedron	1.26	1.523
5	Jupiter	Tetrahedron	3.00	3.416
6	Saturn	Cube	1.73	1.833

Kepler reasoned, there are a maximum of six possible spheres that can be nested among them.

Within the framework of Plato's atomic and Kepler's cosmological models we find a remarkable mathematical synergy (albeit an historically contrived one) between the microcosm and macrocosm. Sadly, perhaps – for they are beautiful ideas – the Platonic description of atomic structure and atomic transformation, along with Kepler's explanation for planetary spacing, are simply wrong. The observations do not support the predictions, and the theories must accordingly be discarded (but not forgotten). Such is the working of science, and scientists must take all such realities within their stride and soldier on. Indeed, just like Plato and Kepler before them, present-day physicists, astronomers, chemists, and mathematicians are still trying to annotate the connections between the very smallest of entities, the atoms, and the largest of all structures, the universe.

A Particle Primer

An outline of the Standard Model of particle physics will be given in Chapter 3. It is a remarkable model, as you will see, but for the moment let us simply look at a few of its key elements.

It has already been stated that all matter is made up of atoms, but it turns out that atoms can be subdivided into even smaller entities – just as Plato allowed the faces of his polyhedral atoms to be subdivided into smaller two-dimensional triangles. The Standard Model describes the essential building blocks of matter and the atom, and it tells us that all stable matter in the universe is made up of just two fundamental particle types: leptons and quarks. There are, in fact, just two leptons and two quarks of interest to the material world. The lepton group is made up of the electron and the electron neutrino, while the quark group is composed of the up quark and the down quark and their antiparticles. All matter that we, as human beings, can see, feel, and experience directly is made up of electrons and combinations of the up and down quarks. The atoms themselves have a centrally concentrated massive nucleus made up of protons and neutrons (each of which is composed of combinations of

Fig. 1.7 The fundamental building blocks of our material world. The nucleus of an atom contains protons and neutrons, and a neutral atom (that is, one with no net charge) will have an associated cloud of electrons

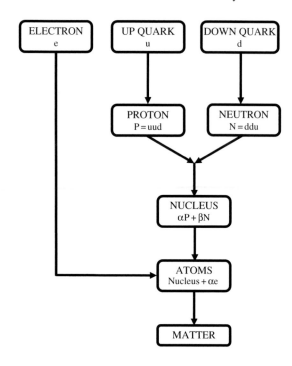

up and down quarks – Fig. 1.7), around which is located a cloud-like structure of electrons.

How all these combinations of quarks, nuclei, electrons, and electron neutrinos and their anti-particles interact will be described in Chapter 3. What follows below is a very brief history of how all this basic understanding came about. It is a remarkable history and one that has been forged by some of the most bright, charismatic, and insightful of beings to have ever lived.

Thomson's Plum Pudding and an Unexpected Rebound

Jumping forward some two and half centuries from the time of Kepler's death, we find ourselves amidst the gentlemanly world of late Victorian science. This was a time of incredible advancement and growing confidence. The world of classical, deterministic physics was at its zenith, and it was beginning to seem that the job of future physicists would soon be reduced to simply finding better and more accurate numbers for experimental constants. The boundaries of physical knowledge had, for so it seemed, been reached.

Nature, however, has a much greater girth than can be encompassed by the iron-clad equations of classical physics, and as the twentieth century approached, the august world of Victorian science was about to be rudely shaken.

The challenge to the systematic rigidity of classical physics came from two fronts. Indeed, it was an overpowering pincer movement, which brought into play experiments relating to the details of atomic structure and observations of the manner in which hot objects radiate energy into space.

The first subatomic particle to be discovered was the negatively charged electron. Building upon earlier experiments relating to cathode rays, Joseph John Thomson, then director of the Cavendish Laboratory at Cambridge University in England, along with his collaborators and students were able to show experimentally, in the late 1890s, that the atom must be able to be subdivided, and that part of its structure was a small, low mass, negatively charged component. These corpuscles, or primordial atoms, as Thomson initially called them, were soon identified with electrons (a name coined by Irish physicist Johnstone Stoney) and associated with the flow of electricity.

Thomson's experiments revealed that the electron mass was very small, and equal to about one one-thousandth the mass of the hydrogen atom. In the wake of Thomson's work a new picture of the atom emerged, and it was reasoned that there must be at least two atomic components – the electrons that carried a negative charge and a confining component that carried a positive charge. Likened by British physicist William Thomson (better known now as Lord Kelvin) to a plum pudding, the image of the atom was that of a positively charged, elemental atomic fluid in the shape of a sphere (the pudding) containing a random distribution of small, negatively charged electrons (the plums).

Ernst Rutherford (Fig. 1.8) was a man with a big, booming voice. Born in New Zealand in 1871, he conducted seminal research at universities in both England and Canada, and while at the Cavendish Laboratory in Cambridge, he acquired the

Fig. 1.8 Ernst Rutherford, 1st Baron Rutherford of Nelson (1871–1937)

nickname "The Crocodile."[1] It was Rutherford who succeeded Thomson as director of the Cavendish in 1919, and it was Rutherford who re-worked Thomson's dense plum-pudding into a more airy and centrally condensed, dare one say soufflé form. Specifically, Rutherford re-defined the positively charged fluid, or pudding, part of Thomson's model.

Rutherford's early research was related to the study of radioactive decay – a phenomenon first described by Antoine Henri Becquerel and the Curie husband and wife team in the late 1890s. While working at McGill University in Montreal, Canada, Rutherford discovered that two types of particles were emitted during the decay of radioactive material: alpha particles, which have a positive charge and weighed in at four times the mass of the hydrogen atom (what we would now call a helium atom nucleus), and beta particles, which in fact are electrons. It was for this pioneering work that Rutherford won the Nobel Prize for Chemistry in 1908.

Further, it was through an experiment devised by Rutherford, while resident at the University of Manchester in England, to study the properties of alpha particles that the nature of the atomic nucleus was revealed. Rutherford didn't actually perform the experiment that is named in his honor; rather, he turned the idea over to his research assistant Hans Geiger (who co-invented the Geiger counter with Rutherford in 1908) and an undergraduate student Ernst Marsden.

The Rutherford scattering experiment (Fig. 1.9) has since become a classic of its kind, and it is now the familiar training apparatus upon which many a present-day atomic physicist has proven his or her experimental mettle. The idea of the experiment is (partly) to see what affect the tightly packed atoms in a gold foil target have upon the direction of flight of alpha particles. If the Thomson plum-pudding model were true, Rutherford argued, then it would be expected that very little scattering away from the central axis would take place, since the electrons, being much less massive than the alpha particles, shouldn't be able to significantly change the latter's path.

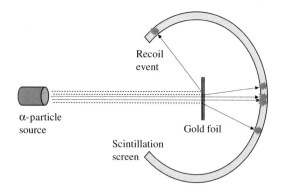

Fig. 1.9 A schematic outline of Rutherford's scattering experiment

---

[1] A crocodile, in veneration of Rutherford, was carved into the wall of the Mond Laboratory building at Cambridge University, during its construction in 1933.

In a completely unexpected fashion, however, what Geiger and Marsden found was that the alpha particles could be scattered through very large angles, and that they could even be back-scattered away from the gold foil. Rutherford summed up the experimental observation thus: "It was truly remarkable. It was as if we fired a 15-inch shell at a piece of tissue paper and it had bounced back."

Rutherford ruminated upon the interpretation of the scattering experiment for nearly 2 years, but in February 1911 he announced what he believed to be the best interpretation of the scattering observations. The atom, Rutherford concluded, must have a massive but exceptionally small central nucleus. With this announcement, the stuffing was literally kicked out of Thomson's plum-pudding, with its positively charged atomic fluid and its randomly distributed sea of electrons being replaced by a small central nucleus about which the electrons swirled.

In some sense the minds-eye picture of the atom that Rutherford presented is similar to that of the planets in our Solar System orbiting the Sun. A Solar System analog picture had, in fact, been proposed before Rutherford announced his interpretation of the scattering experiment by Japanese physicist Hantaro Nagaoka in 1903. Nagaoka, however, likened his atomic model to the planet Saturn, with Saturn itself being the nucleus and its ring system representing the possible orbits of the electrons.

It was through the Rutherford scattering experiment that the present-day picture of the atom emerged. The small, central nucleus of the atom has dimensions of order 10^{-14} m, and this is where the vast majority of the atom's mass resides. The cloud of electrons that swarm about the nucleus is some 10,000 times larger than the nucleus, with a characteristic dimension of about 10^{-10} m. Incredibly, the Rutherford scattering experiment had demonstrated that an atom is composed mostly of empty space.

Further experimentation during the 1920s and 1930s revealed that even the nucleus must have a substructure. There was a positively charged component for which Rutherford eventually coined the name "proton" – after *protos*, meaning first – and then in 1932 one of Rutherford's former doctoral students, James Chadwick, identified the neutron – a neutrally charged particle with a mass almost identical to that of the proton. The different types of atoms that have since been identified, and how they group together to form matter, will be discussed later in this chapter. For the moment, there is a fundamental problem that has yet to be dealt with, that the planetary analog model of the atom, as described by Nagaoka and Rutherford, can't possibly be right.

The Quantum World and the Bohr Atom

"The whole procedure was an act of despair because a theoretical interpretation had to be found at any price." So wrote German physicist Max Planck (Fig. 1.10) in defense of the radical departures from classical mechanics that he had invoked to

Fig. 1.10 A young Karl
Ernst Ludwig Marx (Max)
Planck (1858–1947) in the
lecture theater

explain blackbody radiation. It was the first year of the twentieth century, and classical physics was about to be stood on its head. Planck's desperate act, introduced
in that inaugural year of the new century, was that radiant energy is quantized, only
existing in discreet units of h (Planck's constant = 6.626205×10^{-34} JS). He came
to this startling and entirely unexpected conclusion in order to avoid a problem that
was known as the ultraviolet catastrophe (Fig. 1.11).

The ultraviolet catastrophe was one of those problems that seemed harmless
enough and certainly not one that might usher in a whole new branch of physics. The
problem for the physicists was that the classically developed theory that described
how much energy a hot object should radiate into space per second per unit area (the
energy flux) as a function of wavelength was at complete odds with the experimental observations. Such a situation is not uncommon in science, and since classical
theory could explain the long wavelength energy flux variations (the so-called
Rayleigh-Jeans law) it was mostly assumed that the short wavelength problem,
where less energy flux was observed than predicted, could be whipped into shape
without too much problem. Try as the physicists might, however, the ultraviolet
catastrophe prevailed – theory always predicting a much greater energy flux at
shorter wavelengths than the laboratory experiments revealed.

When Planck decided in the mid-1890s to tackle the problem of blackbody radiators he was eventually forced to abandon classical thinking, such as that which
had resulted in the Rayleigh-Jeans law. Instead he found that he could describe
the experimental results if he assumed that electromagnetic radiation could only be

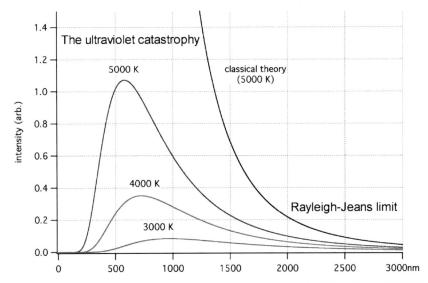

Fig. 1.11 Blackbody radiator curves are shown for temperatures of 3,000, 4,000, and 5,000 K. The classical theory curve (for 5,000 K) predicts an ever larger energy flux at shorter wavelengths. Laboratory observations had shown, however, that the exact opposite happened. This was the ultraviolet (short wavelength) catastrophe that Max Planck solved by introducing the idea of quantized energy

emitted in quantized units of frequency. It was a completely new idea, and his resultant equation for the energy flux was a perfect match to the laboratory-derived data and, unlike the classical theory, it correctly predicted a smaller and smaller energy flux at shorter and shorter wavelengths. It was for this work that Planck received the Nobel Prize for Physics in 1918.

Planck was a reluctant hero, and initially he didn't fully appreciate the incredible importance of his new approach. To Planck the quantization solution was just a mathematical convenience. It was left to other physicists to grab hold of, nurture, and then raise the fledgling theory of quantum mechanics.

The next crucial step in the story was initiated, in fact, by Albert Einstein, who published in 1905 his *anus mirabilis*, a quantum mechanical explanation for the photoelectric effect – work for which Einstein was eventually awarded the Noble Prize for Physics in 1921. Einstein realized that light could be thought of not just as a wave-like phenomenon, as had been shown by James Clerk Maxwell in the 1860s, but as a particle-like phenomenon as well, and accordingly he introduced the idea of the photon, which can be thought of as a localized packet (that is particle-like) burst of an electromagnetic wave. Although Einstein later came to reject many of the philosophical ideas relating to quantum mechanics, his explanation for why electrons could be knocked from certain metals by shining a light upon them indicated that Planck's radical idea was not just a one-off fix for blackbody radiators.

It was not long after the pioneering research papers by Planck and Einstein appeared that other physicists started to look for additional connections between the quantum hypothesis and the observed behavior of atoms – especially atoms in a heated gas. The crucial linkage was eventually forged by Danish physicist Niels Bohr (Fig. 1.12) in 1913, then working under the guidance of Ernst Rutherford at the University of Manchester.

Fig. 1.12 Niels Henrik David Bohr (1885–1962) in the lecture theater

Starting with the ideas developed by Rutherford, Bohr envisioned the atom to be composed of a massive, positively charged nucleus about which orbited negatively charged electrons. Importantly, however, and this was Bohr's key innovation, the energy associated with the various electron orbitals was quantized. Electrons, counter to the dictates of classical theory, couldn't occupy just any orbit but only those orbits that were allowed electron orbits for the specific atomic nucleus being considered.

By insisting that the electrons could only occupy very special, allowed orbits, Bohr solved the existence problem that Nagaoka and Rutherford had inadvertently introduced with their planetary models of atomic structure. The problem for the planetary model was that it placed no constraint upon the energies associated with the allowed electron orbits; they could have any value, and since a charged particle moving along a curved path must emit radiation, a dictate that results from James Maxwell's theory of electromagnetic radiation, so the electron must lose energy, spiral inwards, and crash into the nucleus.

The electron orbits in the planetary model were not stable since the electrons must continuously radiate away their orbital energy. In the quantized Bohr atom, however, an electron can only move from one allowed electron orbit to another if it gains or loses the exact amount of energy to enable a transition. In essence the quantum model is an all or nothing energy game that the electron plays. Either the energy gained or lost by the electron is exactly right to enable a transition, or it stays where it is. The inspirational step that Bohr took in order to make Rutherford's planetary model work was to rewrite the rules concerning the allowed behavior of the electrons and to describe exactly which electron orbits are permitted. For his fundamental contributions to the understanding of the atom Bohr was awarded the Nobel Prize for Physics in 1922.

Within the macrocosm is the microcosm, and accordingly it was through the study of the emission lines associated with a heated hydrogen gas that Niels Bohr was able to fully describe the energy levels associated with the hydrogen atom, and thereafter, that astronomers were able to classify the stars. Key to both of these great innovations was the observation made by Swiss school teacher Johann Jakob Balmer (1825–1898).

Spectroscopy is the study of light emission and absorption. Specifically, it is concerned with quantifying how a liquid, solid, or gas either emits or absorbs electromagnetic radiation according to wavelength. A pure hydrogen gas in a thin glass tube, for example, will emit very specific wavelengths of light when an electric current is made to run between the ends of the tube. It was in 1885 that the 60-year-old Johann Balmer first studied and quantified the specific wavelengths of the light emitted by an excited hydrogen gas. He found that there was an emission line corresponding to red light at a wavelength of 6.5621×10^{-7} m, other emission lines at shorter wavelengths corresponding to the blue-green line at 4.8608×10^{-7} m, and the violet-colored emission line at 4.340×10^{-7} m.

The question, of course, is why is there emission of light at these very specific wavelengths, rather than emission at all wavelengths? The reason that we see discrete rather than continuous radiation is beautifully explained by Bohr's quantized orbital model (Fig. 1.13). A two-stage process is required for such emission lines to come about. First the atoms in the gas must be excited, and this is achieved by heating the gas. In this manner the atoms begin to move around at high speed, and collisions between various pairs of atoms are capable of providing just the right amount of energy for an electron to be placed into a high energy orbital.

Once situated in a high energy orbital, however, the electron can spontaneously jump down to a lower energy orbital by emitting a photon that carries away the excess energy. This process is illustrated in Fig. 1.13, where an electron is shown making a transition from the $n = 3$ orbital to the $n = 1$ (ground state) orbital. The energy carried away by the photon will be exactly equal to the energy difference between the starting and ending orbitals. The light from such a transition will always have the same wavelength, because the electron orbital energies are fixed, and such light is said to be in emission because it is the photon that carries away the energy. Transitions to and from the $n = 1$ ground state of hydrogen were first studied by American physicist Theodore Lyman in the first decade of the twentieth century, and accordingly they are often called Lyman lines.

The reverse process can also work, of course, with the electron absorbing energy to jump from the ground state to the $n = 3$ orbital. Since such a jump requires the electron to gain energy it is called an absorption transition, and the energy of the photon that enables such a transition will be exactly the same as the energy carried away by the photon in the reverse, emission process.

The series of hydrogen lines studied by Balmer, it turns out, are all associated by the fact that they involve transitions in which the electron jumps up from

Fig. 1.13 A schematic
outline of the hydrogen atom
and its first three allowed
electron orbits $n = 1, 2, 3,\ldots$.
The first two Lyman
absorption lines emanating
from the ground state ($n = 1$
to $n = 2$ and $n = 1$ to $n = 3$)
are shown and so, too, is the
first Balmer absorption line
emanating from the second
allowed electron orbit (the
$n = 2$ to $n = 3$ transition).
The $n = 3$ to $n = 1$ Lyman
emission line is also
illustrated

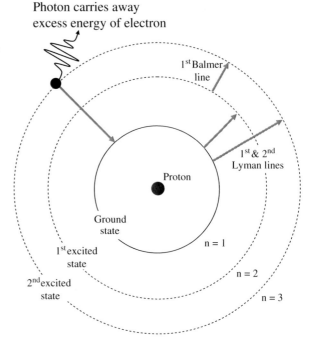

or down to the $n = 2$ (second allowed) orbital. Indeed, there is a remarkably simple formula that describes the wavelength of the various Balmer lines. If the transition takes the electron from the mth orbital to the $n = 2$ orbital, then the wavelength of the emitted photon will be: $\lambda_m = B/(1 - 4/m^2)$, where B is a constant equal to 3.6456×10^{-7} m. The first Blamer line [also called the hydrogen alpha (Hα) line] corresponds to the $m = 3$ to $n = 2$ transition and has, according to our formula, a wavelength corresponding to $\lambda_3 = 9B/5 = 6.5621 \times 10^{-7}$ m. It was by deriving Balmer's empirical formula in terms of the quantized electron orbitals that Bohr knew he was on the right track, and that the underlying relationship between the spectroscopic observations and atomic structure had been explained.

Long before Niels Bohr had explained the quantum process at work in the production of Balmer lines, astronomers had been using them to classify stars. The work began in the last quarter of the nineteenth century, and the Balmer lines were chosen as a diagnostic feature simply because they were prominent and fell in the visible part of the electromagnetic spectrum.

It is only within the past few decades that astronomers have been able to study emission and/or absorption lines in the infrared, ultraviolet (at which wavelengths the Lyman lines are to be found), and X-ray regions. Stars generally show a composite spectrum that is derived from a combination of a blackbody curve (recall Fig. 1.11) and wavelength specific absorption lines. The blackbody curve is produced by the high-density, high-temperature interior gas, while the absorption lines are produced through the presence of atoms located within the outer, cooler, less-dense stellar envelope. In the case of the Sun, the absorption lines are all produced within a photosphere that is just a few hundred kilometers in thickness – a zone that constitutes about 0.05% of its radius.

Even though the region of the photosphere is veneer thin, the absorption lines produced there enable astronomers to learn an incredible amount about the underlying star. If stars were perfect blackbody radiators we would be able to determine their temperatures, but nothing else. By showing absorption lines, however, astronomers can determine the composition of a star, and they can measure its rotation rate as well as determine if it has a magnetic field or if it is losing mass into interstellar space. Within the macrocosm is the microcosm, and what astronomers now know about the stars has been revealed by understanding the manner in which electrons move between their allowed electron orbits in those atoms located within the outermost layers of stellar atmospheres.

When astronomers first began to study stellar spectra in the later part of the nineteenth century it was not clear how to interpret their physical meaning. With the introduction of quantum mechanics, however, it became possible to show that the strength of an absorption line was determined by the temperature of the star and the abundance of the atomic species present. In this manner astronomers were able to deduce that stars are predominantly made of hydrogen and helium, with just a smattering of all other elements.

If one analyzed the compositional makeup of a 100-g clump of solar material, for example, then 71 g of it would be in the form of hydrogen, 27.1 g in the form of helium, and all other elements would make up the remaining 1.9 g. The analysis of other stars reveals a very similar picture; they are all predominantly composed of hydrogen and helium with a small smattering of other elements. Intriguingly, the further back in time that a star actually formed, that is, the older it is, the fewer elements other than hydrogen and helium it is found to contain. This observation, as we shall see later in this chapter, tells us something about the very beginnings of the universe, and it also betrays the existence of an incredible matter recycling program at work within our galaxy.

The New Quantum Mechanics

Einstein's idea that light can be thought of as having both wavelike and particle-like characteristics was truly radical, and it stretches the imagination to near breaking point. We can easily imagine a wave, and we can easily imagine a particle, but what does a wave-particle look like? Understanding the meaning of this so-called wave-particle duality was central to the development of quantum mechanics, and it was also pivotal in the demise of the classical, deterministic view of atomic structure. Although the Rutherford-Bohr planetary-atom picture (Fig. 1.13) provides a correct explanation for the emission line wavelengths in the hydrogen atom, it is nonetheless far too simplistic a view of what the hydrogen atom actually is and how the electrons truly behave.

The new quantum mechanics emerged amidst the shattered and radicalized world of post-World War I Europe. This was the time of the Dada arts movement, Bauhaus architecture, the new psychology of Carl Jung, the politics of Marxism, and the modernist plays of Bertold Brecht. All was in an intellectual flux, and every facet of established wisdom was opened up to question, reinterpretation, and new ways of thinking. French physicist, prince, and later Duke Louis Victor de Broglie ushered in the new quantum era with the publication of his doctoral thesis in 1924.

If light, he reasoned, long thought of as a wave phenomenon, can have particle-like characteristics, as argued by Einstein, then why can't a particle, such as the electron, have wave-like characteristics? It is a reasonable enough thought, but if it is true then the consequences for physics and our interpretation of the atomic world are profound. Within the pages of his Ph. D. thesis, de Broglie argued that the wave-like properties of a particle can be described according to a characteristic spatial distance or wavelength, a distance now known as the de Broglie wavelength λ_B, such that $\lambda_B = h/p$ where h is Planck's constant and p is the particle's momentum. In the classical world, linear momentum is the product of an object's mass and velocity, while the angular momentum of an object moving in a circular orbit is defined as the product of its mass, velocity, and orbital radius. It was by forcing the angular momentum of the electron to be quantized that precipitated Niels Bohr's breakthrough in understanding the hydrogen atom. The remarkable and very strange to our everyday senses implication of de Broglie's argument, however, is that the greater a particle's momentum (say, the faster it is moving), the smaller is its associated wavelength. Although this idea may seem at odds with our view of the everyday world, it is in fact this very action that allows electron microscopes to work and collider experiments (such as those that will be conducted in the LHC) to probe the innermost structure of the atomic nucleus – as we shall see.

Werner Heisenberg (Fig. 1.14) was born in the year that Max Planck published his revolutionary paper on blackbody radiators. By the relatively tender age of 25 years, however, Heisenberg was performing post doctoral research work at the Niels Bohr Institute in Copenhagen, Denmark. His findings there would change physics and human perceptions about the world forever. What Heisenberg realized was that there is a limit to how much we can know about certain physical quantities. We cannot, for example, measure both the position and momentum of a particle to any arbitrary precision. Our view of the quantum world is limited; the more certain we are of a particle's position, for example, so the less certain is our knowledge of its momentum, and vice versa. If the uncertainty in the position measurement is written as Δx and the uncertainty in the momentum measurement is written as Δp, then Heisenberg showed that the product $\Delta x\, \Delta p$ must always be greater than $\hbar/2$, where $\hbar$ is the standard symbol for Planck's constant divided by 2π. Within this famous equation, indeed an equation almost as famous as Einstein's $E = mc^2$, lies the death of classical determinism.

Fig. 1.14 Werner Heisenberg (1901–1976)

Heisenberg's uncertainty principle can be expressed in terms of what is called the wave function – a mathematical expression that contains all of the important wave-like information about the position and momentum of a particle. In this picture, the uncertainty of the quantum world arises when an observation is made, and the wave function is said to be collapsed. If, for example, a highly accurate measurement of a particle's position is made, meaning that the part of the wave function describing the particle's position collapses to a small region of space pinpointing its location, so the uncertainty condition requires that the component of the wave function describing the particle's momentum spreads out, becoming non-localized, and accordingly our knowledge of its actual value becomes unclear. In our everyday macroscopic world the uncertainty principle is far too small to have any directly observable consequences, but in the quantum realm its rule is absolute. In spite of the great rallying cry of the Greek philosopher Protagoras of Abdera, man, after all, is not the measure of all things – at least simultaneously, that is.

The detailed properties of the wave function need not concern us here, but it is worth mentioning that in 1926 Austrian physicist Erwin Schrödinger was able to show that it can be thought of in terms of probability amplitudes for the different possible configurations that a quantum system can be in. Schrödinger's famous wave equation further describes the time evolution of the probability of finding a particle at a specific location. In this manner the rules and equations of quantum mechanics do not provide us with an exact indication of a particle's position at every moment, but only a probability of where it is likely to be at any specific instant.

Exclusion

Austrian physicist Wolfgang Pauli (Fig. 1.15) was a great letter writer, but he rarely published in scientific journals. His key contribution to our story, however, dates

Fig. 1.15 Wolfgang Pauli
(1900–1958)

to a 1925 research paper in which he outlined why different kinds of atoms can exist and why, in fact, chemistry works. The reason, of course, is quantum mechanics and specifically a non-overcrowding condition that electrons must abide by. The so-called Pauli Exclusion Principle (PEP) states than no two identical electrons (or fermions specifically – see Chapter 3) can occupy the same quantum state simultaneously.

For the electrons that surround an atomic nucleus the PEP explains why they are not all located in the lowest energy state but are spread out to occupy the shell structure described by spectroscopists. The PEP also explains the manner in which atoms can share electrons and thereby form composite (that is, molecular) entities. All matter, ourselves included, have bulk and remain stable against collapse because of the PEP – a remarkable result and one that once again underpins the point that within the very large is the very small. If it wasn't for the PEP in the microcosm, there would not be a macrocosm – period.

Fermi's Little Neutron

The neutrino is a ghost of a particle, and as is the nature of a ghost so the neutrino can pass through solid walls and change its appearance from one spectral form to another. A shape-shifting, antisocial particle if ever there was one, and remarkably, the existence of the neutrino was predicted long before it was possible to experimentally detect it. That the neutrino must exist was predicted by Wolfgang Pauli in 1930, and the idea was predicated upon an absolute faith in the concept of the conservation laws of nature.

The charismatic and now sadly missed American physicist Richard Feynmann once eulogized that the discovery of the conservation law of energy was one of the supreme triumphs of science. Indeed, it is an absolute and inviolate rule: within a closed system isolated from external influence the total amount of energy is always constant. There is no deviation from this rule. It is not an approximate result, right some of the time and wrong at others; it is an absolute rule, and there are very few such rules in all of nature. This all being said, as we shall see Chapter 7, the mind-bending world of quantum mechanics does actually allow for fleeting deviations of this fundamental rule, a result that follows from Heisenberg's uncertainty principle. Energy can certainly change its form in a closed system, from that of motion, for example, to that of sound and heat after a collision. But the sum total of all the various energy terms must always add up to the same number. It was by such reasoning that the existence of the lightweight, neutrally charged neutrino was correctly predicted.

The name neutrino was coined by Italian-American physicist Enrico Fermi (Fig. 1.16) and is derived from the Italian word for the neutron – *neutrone*. In Italian, this word can be given a diminutive sense by changing the ending *one* to *ino*, and accordingly the newly predicted low-mass, neutral particle became the little neutron or neutrino (so work the philological minds of physicists). Pauli, along with Fermi and numerous other researchers, realized that this new particle, the neutrino, must exist after studying in detail the process known as beta decay. In beta decay a nucleus spits out a beta particle (an electron or position). One example of the beta decay process is that in which a free neutron (N) decays into a proton (P), an electron (e), and a neutrino (ν).

Fig. 1.16 Enrico Fermi. This now legendary photograph shows Fermi at the chalkboard along with an incorrectly written formula for the fine structure constant α, proving that even Nobel Prize-winning physicists are human after all. The formula should read $\alpha = e^2/\hbar\,c$. The other equation shown on the chalkboard is Schrödinger's equation for an atom with Z protons in its nucleus (the hydrogen atom corresponds to $Z = 1$)

That the neutrino had to exist became clear when the energy budget of the beta decay process was analyzed in detail. Just as an accountant at a bank will carefully match the funds being paid out to those being paid in, so the experimental physicists found that following the beta decay of a neutron the energies imparted to the proton and electron were always less than expected – there was an energy deficit, and energy, it was known, couldn't just disappear. The answer to this energy deficit problem, however, can be solved by invoking the existence of an additional particle, the fleet-footed neutrino, to make up the energy total. Accordingly, the beta decay of the neutron is written in full as $N \Rightarrow P + e + \nu$, where ν is the mysterious neutrino.

The decay process of the neutron is actually mitigated by the weak interaction that will be described in Chapter 3. The first direct experimental detection of neutrinos was made in 1956, nearly 30 years after they were predicted to exist. It is now known that there are three neutrino types (called flavors) and their associated anti-particles.

By far the largest nearby source of neutrinos to us is the Sun. Within its core, where hydrogen is fused to helium (a process that will be describe shortly), a veritable sea of neutrinos is produced – indeed, about 2×10^{38} electron neutrinos are produced within the Sun every second. In Earth's orbit this neutrino production rate results in a flux of 7×10^{14} neutrinos/m^2/s. In the minute required to read this page something like 40,000 million million solar neutrinos will pass through your body.

Given the great abundance of solar neutrinos passing through every square meter per second at Earth's surface one might think that detecting them would be easy. The reality of the situation, however, is far more complex. Neutrinos are almost perfect ghost particles, and only very, very, very rarely interact with the corporal world. Undaunted by such prospects, however, American chemist Ray Davis (Brookhaven National Laboratory) set out in the late 1960s to look for the diaphanous traces of solar neutrinos.

To begin with, Davis headed nearly a kilometer and half underground. To observe solar neutrinos one must first shield the detector from as many background sources of interference (such as cosmic rays) as possible. To the neutrino, however, a kilometer or so of solid rock is completely invisible. It was within a vast chamber (Fig. 1.17) in an unused part of the Homestake Gold Mine in South Dakota that Davis and co-workers set up a vast tank containing 455,000 liters of tetrachloroethylene (a dry cleaning fluid). The reason for using a dry cleaning fluid was that it provided a ready supply of chlorine atoms, and these were the targets for the solar neutrinos. It turns out that a neutrino can interact with a chlorine atom to produce argon, and the idea that Ray Davis came up with was to flush the tank once a month and in the process search out how many argon atoms had been produced.

This all sounds straightforward, but for a typical month-long exposure time, perhaps 10 argon atoms might be produced, and detecting these few atoms would be a formidable task. Davis and his team of researchers, however, managed to detect the presence of newly generated argon atoms within the holding tank, indicating that neutrino interactions must have taken place. There was a problem, however; only about one third of the expected number of neutrino events were being recorded. Either the experiment was not behaving as expected, or the theoretical models

Fig. 1.17 The large tank containing 455,000 liters of dry cleaning fluid – the liquid heart of the Homestake Mine solar neutrino experiment. (Image courtesy of the Brookhaven National Laboratory)

describing the Sun's energy generation reactions were wrong (or both). Time passed; the theoreticians worked on their models, and the predicted neutrino flux changed, but very little; it was the experiment that had to be wrong. Davis, being a skilled experimenter, however, would have none of it, and after exhaustive testing and tweaking he insisted that the detector was working perfectly and that the observed flux was, in fact, a third of the predicted one.

An impasse between the theoreticians and the experimentalists emerged, with neither side giving ground. In the end, it turned out that both groups were right. Davis was counting the correct number of neutrinos, and the theoreticians had predicted the correct total number. What emerged, following nearly 25 years of wrangling, was that the properties of neutrinos had not been fully understood.

The Homestake neutrino detector designed by Davis and his collaborators was sensitive to only one specific type of neutrino, the electron neutrino, and this, it turned out, was the root of the problem. In spite of initial expectations it turns out that neutrinos actually have a small associated mass, and this means that they undergo so-called flavor oscillations, such that as time goes by the neutrino shuffles between the three various allowed types. This is why the Davis experiment only measured one-third of the predicted flux, since it couldn't detect the other two types of neutrino. The Sudbury Neutrino Observatory (SNO), located in a deep mineshaft in Sudbury, Canada, in 2002, beautifully confirmed this remarkable picture of neutrino flavor-changing behavior. The SNO detector uses 1,000 tons of heavy water, rather than a dry cleaning fluid, for its target medium, and it can detect all neutrino flavors. After just a few years of operation, the definitive SNO results were in, and the neutrino count was exactly as predicted by the solar models. The solar neutrino problem had been solved, but there are still many unknowns associated with the neutrino.

A number of neutrino experiments are currently being accommodated at CERN, and the Super Positron Synchrotron (to be described in Chapter 2) provides a direct beam of neutrinos that travels over 700 km underground (no tunnel is actually required, you'll recall, as the neutrinos will happily sail straight through the rock) to the Gran Sasso National Laboratory in central Italy (Fig. 1.18), where the experiments are actually performed.

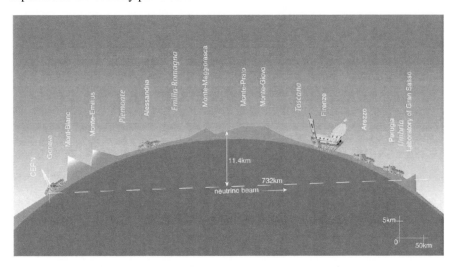

Fig. 1.18 The CERN neutrino beam line. The various experimental halls at the Gran Sasso National Laboratory are located underneath the massive dolomite shield provided by the Apennine mountain range. (Image courtesy of CERN)

It takes just 2.5 ms for the neutrinos to travel from CERN to Gran Sasso, where the OPERA (Oscillation Project with Emulsion-tRacking Apparatus) waits in the hope of capturing a photographic snapshot of the destruction of a few of them. OPERA is a massive and essentially neutrino-stopping brick wall. The experiment is composed, in fact, of about 150,000 photographic bricks. Each brick weighs in at about 8 kg and is made up of 57 layers of photographic film interleaved between thin lead plates. The photographic blocks will provide a diagnostic three-dimensional track record of any neutrino interactions that chance to occur within the experiment, and the hope is to specifically capture so-called tau-neutrino decays (see Chapter 3). Since the CERN beam is composed of muon neutrinos only, the tau-neutrino events must be derived as a consequence of flavor oscillations – and this is the main focus of the study.

Three Quarks for Muster Mark!

The discovery that both protons and neutrons, the particles of the atomic nucleus, are not elemental but composed of even smaller entities became clear as a result of collider experiments conducted during the 1950s and 1960s. These findings beg the question, how is it possible to probe the atomic nucleus when it is so very small?

The answer lies in the idea developed by Louis de Broglie (introduced earlier). The more energy that is imparted to a given particle, say an electron, the greater is its momentum, and accordingly the smaller is its de Broglie wavelength. The structure of protons and neutrons can be studied, therefore, by firing very high velocity electrons at them. To probe atomic scales with dimensions of 10^{-15} m, the electrons must be accelerated to energies of order 1 GeV ($\sim 1.6 \times 10^{-10}$ J).

Although these energies place us in the realm of high-energy physics they are entirely paltry to the everyday world around us. The common housefly (*Musca Domestica*) weighs in at about 12 mg in mass and can reach a top speed of about 2 m/s. The housefly, therefore, in just flying around a room carries 200,000 times more energy (that is, about 3×10^{-5} J) than our withering electron projectile. The reason why electrons can probe atoms and houseflies can't, of course, is due to the concentration of the energy into a small region. Our 1 GeV collider electron delivers its energy to a region that is 10 million million times smaller than the body of our buzzing fly.

Once it became possible to accelerate charged particles, not just electrons, to very high velocities, physicists performed Rutherford-style scattering experiments. The scattering now, however, probed the structure of the nucleus rather than the entire atom. By 1969 the state of play was such that it was clear that protons and neutrons must harbor smaller-scale structures. Not only this, as colliders became more and more powerful, more and more short-lived collisional decay particles appeared; indeed, one could even say an embarrassment of new particles were detected. Theoretical physics was in crisis, and a revised fundamental theory of atomic structure had to be found. Not only did the new theory have to explain what the structural components of the proton and neutron were, it also had to explain and classify the panoply of other particles that were being detected.

Many physicists were working on the problem, but the first person (although to be fair one should say with the help of many collaborators) to hit upon what is now accepted as the right explanation was Murray Gell-Mann (the California Institute of Technology) in 1956 (Fig. 1.19). Gell-Mann argued that there were three fundamental particles, which he eventually called quarks (after a passage in James Joyce's rather arcane novel, *Finnegan's Wake*). The quarks themselves were assigned fractional electric charges, but since no such charge effect is ever observed, this required that quarks must always be found in groups. The three quarks were given the names up (u), down (d), and strange (s), and their electrical charges were assigned as +2/3, −1/3, and −1/3, respectively.

The proton is made up of two up quarks and a down quark (written as uud) and accordingly has a positive +1 charge. The neutron is composed of one up quark and two down quarks (written udd) and accordingly has zero charge. How the quarks

Fig. 1.19 Murray Gell-Mann. (Image courtesy of the Santa Fe Institute)

fit together to explain the existence of other particles will be outlined in Chapter 3. Gell-Mann was awarded the Nobel Prize for Physics in 1969 as a result of his work relating to the identification and understanding of the fractionally charged quarks.

We have now completed our brief survey of the Standard Model. The scale explored so far is illustrated in Fig. 1.20. Further details of the many additional subatomic particles will be given in Chapter 3, along with a discussion of what new

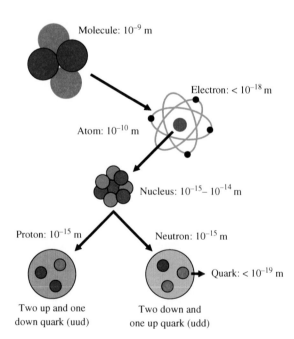

Fig. 1.20 The characteristic sizes of the particles (electrons, protons, neutrons, and quarks) that constitute the atomic world

physics might exist beyond the Standard Model. In the remainder of this chapter, our focus will shift to how atoms form matter, and where atoms themselves came from.

Building the Universe

Our everyday world is made up of countless trillions upon trillions upon trillions of atoms. Indeed, atoms are the microcosm to our macrocosm; they are the components out of which we, and everything around us, are made. And, just as the atomic nucleus has structure and order, so, too, do the greater collections of atoms; they interact in special and very specific ways, and chemistry is the science that makes sense of the processes that allow atoms to interact and form molecules. From the smallest scale to the largest scale, atoms combine into molecules and molecules embrace each other to make larger structures, and so eventually the world we experience is brought into a tangible form.

Remarkably, indeed incredibly, some of the immense groupings of atoms and molecules that exist in the universe are animate and alive, and if they are human they might even ask questions about their origins. Such human entities can also look for order and structure amidst the great blooming, buzzing confusion of their existence.

The Matter Alphabet

Just as all words are made up of letters taken from the alphabet, so, too, is matter made up of atoms selected from the "matter alphabet." In addition, just as the English alphabet running from A to Z is finite, containing just 26 symbols, so, too, is the matter alphabet composed of a finite list of stable atoms.

The matter alphabet, or Periodic Table as it is more commonly called, contains 94 natural elements and about 20 additional laboratory-synthesized elements not found in nature. Indeed, the term "natural" is employed here to indicate that the elements are found on and inside Earth. All atoms beyond 94 in the matter alphabet are unstable, with the most extreme heavyweights being synthetically produced in particle accelerators, their lifetimes typically being measured in billionths of a second.

The current record holder among the synthesized elements takes the matter alphabet to letter 112. This fleeting heavyweight was first produced in 1996, but only officially added to the Periodic Table in June 2009. With its official status recognized this latest record-breaking element was named copernicum, in honor of astronomer Nicolaus Copernicus (whom we shall meet in Chapter 4).

The matter alphabet is arranged in a very specific order, according to the number of protons in the atomic nucleus. The first letter of the alphabet therefore is hydrogen (H), which consists of just one proton in the nucleus. The second letter is helium

(He), consisting of two protons and two neutrons in the nucleus. The 94th letter in the natural matter alphabet is plutonium (Pu), which contains a whopping 94 protons and 150 neutrons in its nucleus.

When forming the equivalent of words and producing sentences from the matter alphabet there are many rules relating to the allowed grammar and letter associations. The field of chemistry is intimately concerned with the structure of matter sentences, and in the understanding of how to write proper matter-words. A gas composed entirely of hydrogen, for example, might be written as the sentence: $(H)_1 + (H)_2 + (H)_3 + \ldots + (H)_n$, where the subscript identifies each atom (all of which can move around freely), and where there are n atoms (usually an extremely large number) in the gas. Our H-gas representation is not exactly Shakespeare, for sure, but it is a correct and decently enough written matter-word. A cloud of pure water vapor might further be described by the sentence: $(H_2O)_1 + (H_2O)_2 + (H_2O)_3 + \ldots + (H_2O)_n$, where we now have a gas of n molecules, each of which is composed of two hydrogen (H) atoms bonded to one oxygen (O) atom.

A pure water ice crystal, in contrast, might be written as the word: $6(H_2O)$, with the 6 indicating that the H_2O molecules are formed into a hexagonal ring structure. The solid ice cube, which one might place in a Martini drink at the end of a long working day, is composed of p repeated layers of an $n \times m$ sheet of multiply linked $6(H_2O)$ crystals. The molecule sodium ethyl xanthate (SEX), commonly used in the extraction of metal sulfides from mining slurry pits, is written according to the matter alphabet as $C_3H_5NaO_2S$, indicating that it is made up of 3 carbon (C) atoms, 5 hydrogen (H) atoms, a sodium (Na) atom, two oxygen (O) atoms, and one sulfur (S) atom. And, when many such molecules all join together in a solid crystalline mass, what do you get? Apparently, for so the chemists tell us, a pale yellow amorphous powder that has a very disagreeable smell.

The entire chemistry of life – liquids, solids, and gases – is constructed from the complex, many-layered matter-words that can be constructed from the matter alphabet. The world is literally writ large from the matter alphabet. Some of the matter-words are actually cast in stone and may be billions of years old; other matter-words appear in the fleeting verse that represents the pleasant aroma of a perfumed vapor.

We Are of the Stars

Although the matter alphabet (or the Periodic Table of Elements, if you prefer) tells us what all the things that we can feel, touch, and smell are made of, it does not tell us where or how the individual letters of the alphabet formed. For indeed, not all of the letters in the matter alphabet are of the same age; some are much older than others, and some are much more abundant than others. Hydrogen and helium, the first two letters of the matter alphabet, are by far the oldest and the most abundant entities in the matter alphabet; they are indeed the originals that formed, as we shall see in Chapter 4, during the first few minutes of our universe coming into existence. Some of the hydrogen and helium nuclei in the matter alphabet are a humbling 14 billion years old. Likewise, the lithium, beryllium, and boron, letters three, four, and five, are of this same great ancestry. All of the other letters, however, are of a much more recent genesis, and rather than being formed in the oppressive heat of primordial nucleosynthesis, the nuclei from letter six (corresponding to carbon) onwards were forged within the crucibles that were once bright and massive stars.

It is a remarkable story. Our very bodies, the very Earth, along with the planets and the Sun, are all made up of atoms and molecules that are both ancient and modern. The hydrogen atoms within the hemoglobin molecule that transports oxygen around our bodies dates to the very beginnings of the universe, but the iron atom (letter 26 in the matter alphabet) that it contains was quite possibly formed within the past 5 or 6 billion years in the blinding explosion of a supernova. The oxygen atoms (letter 8 in the matter alphabet) that the hemoglobin molecule transports were also forged within the interiors of now long-dead stars. We are literally alive in the here and now because of the journey work of the stars and their far flung ashes.

The entities that we call stars are essentially energy conversion machines. They start off with only potential gravitational energy, and their entire existence is dedicated to one shorter or longer, according to initial mass, attempt at stopping gravity from crushing them out of existence. Indeed, if nothing opposed the gravitational force that entwines our Sun, it would collapse down to nothingness (forming, in fact, a black hole) in about 50 min. What holds gravity at bay, and enables the Sun to shine – and ultimately allows us to exist – is the conversion of one atomic species into another via fusion reactions. It is by the transmutation of one letter in the matter alphabet to another letter that enables the matter alphabet to grow, but this transformation process only works in regions where the density and temperature are very high. At the Sun's center, for example, the density is about 13 times greater than that of lead, and the temperature is a blistering 15 million degrees Kelvin. Under these hellish conditions the proton–proton chain of fusion reactions allows for the conversion of four hydrogen atoms into a helium atom and energy.

Schematically, in matter letters, we can write $4H \Rightarrow He +$ energy. This basic reaction doesn't actually extend the matter alphabet, since helium has existed since the very earliest of moments of the universe. Importantly, however, it is the energy generated in the conversion of hydrogen into helium that keeps the Sun from collapsing. Indeed, the energy generated by fusion reactions at the center of the Sun exactly balances the energy lost into space at its surface, and the temperature gradient setup within the Sun is able to maintain the pressure balance needed to support the weight of overlying layers at every point within its interior, and thereby gravity is kept in check.

Stars converting hydrogen into helium via fusion reactions are said to be main sequence stars, and this is their longest-lived stable stage. The Sun, for example, will be able to tap its reserves of hydrogen fuel for about another 5 billion years (it is already 4.5 billion years old). Importantly, however, the main sequence lifetime of a star varies inversely with its mass. The more massive a star the shorter is its main sequence phase. A star 30 times the mass of the Sun (the evolution of which is described in Chapter 7) has a main sequence lifetime of just a few million years – a literal blink of the eye even on a geological timescale.

With the conversion of all its central hydrogen into helium a star is left wanting for an energy source. Something will have to give, or the star will cool off and collapse under the withering stranglehold of its own gravity. What eventually happens is a new alchemy. With the exhaustion of hydrogen the central temperature and density of a star climb to new heights of intensity; the temperature approaches 100 million Kelvin, and the density increases to an oppressive 10,000 times that of lead.

Finally, within the confines of this crushing inferno, the first new letter of the matter alphabet is formed. What happens is that the triple-alpha reaction kicks in, and three helium atoms combine to produce a carbon atom (the fifth letter in the matter alphabet). Schematically, we can write 3He $\Rightarrow$ C + energy. With the onset of the triple-alpha reaction a star, once again, has an internal energy supply that can replenish the energy it loses into space at its surface, allowing it to stave off a catastrophic gravitational collapse.

The stellar readjustments that follow the onset of helium burning result in the star swelling up to gigantic proportions, and it becomes a red giant. The prefix red is used to describe these stars, since they have low surface temperatures, perhaps just 3,000 K (half that of the present-day Sun), and accordingly, from Wien's law relating to blackbody radiators, most of its energy is radiated at wavelengths corresponding to red light. If you take a look at the star Betelgeuse, the brightest star in the constellation of Orion, the next time it is above your horizon at night time, you will see that it has a distinctly reddish color to it. Betelgeuse is a helium-burning red giant star with a radius about 950 times greater than that of our Sun (Fig. 1.21).

Within the central cores of red giant stars helium is steadily converted into carbon, and the matter alphabet is expanded beyond the first five elements that were produced in the primordial nucleosynthesis of the Big Bang (to be discussed in Chapter 4). But the stars are far from being finished, and the matter alphabet has many more elements that need to be forged.

What happens after the red giant phase depends upon the original mass of the star. Detailed numerical calculations indicate that the critical mass is set at about eight times that of our Sun. Above eight solar masses a star can undergo advanced stages of nuclear processing within its core. Below eight solar masses and the red giant, helium-burning phase is as far as the nuclear processing can go; the ultimate end phase for such stars is a white dwarf. White dwarfs are near pure carbon spheres supported by the non-overcrowding pressure exerted by degenerate electrons obeying the Pauli Exclusion Principle. We shall have more to say about white dwarfs in Chapter 6.

Those stars with initial masses greater than eight times that of our Sun burn brightly, die young, and leave a disintegrated corpse. It is through their explosive death throes that the matter alphabet is pushed to its furthest limits. All atoms beyond carbon are synthesized in massive stars through advanced nucleosynthesis reactions before and during the final supernova phase triggered by the catastrophic

Fig. 1.21 The red giant star Betelgeuse, the brightest star in the constellation of Orion. If suddenly switched with our Sun, Betelgeuse would fill the Solar System outward to the orbit of Jupiter. (Image courtesy of NASA)

and very rapid collapse of their cores. The force of gravity ultimately wins, and its chance to reduce a massive star to nearly nothing arrives with the synthesis of iron.

Once carbon has been produced by the triple-alpha reaction it can then interact with another helium atom to produce oxygen (letter 8 in the matter alphabet). Two carbon atoms can also combine to produce neon (Ne – letter 10) or sodium (Na – letter 11). Once sufficient oxygen has been synthesized then silicon (Si), phosphorus (P), and sulfur (S) can be produced. Finally, at least for the star, reactions with silicon can result in the formation of atoms up to iron (Fe –the 26th letter in the matter alphabet).

Since each of these successive nuclear reactions require ever higher densities and temperatures to operate, the interior of a star ends up with a layered structure, rather like that of an onion (Fig. 1.22). At the very center is a small iron core (with a mass about equal to that of our Sun but a size equivalent to that of Earth). Surrounding the iron core is a shell rich in silicon, surrounding that is a shell rich in oxygen, and so on outwards until the original hydrogen- and helium-rich outermost envelope is encountered.

With the development of an iron core the star has only a very short time left to live. The problem with the formation of iron is that it cannot be converted into any other element with the associated release of energy. Remember it has been the release of energy in all the other fusion reactions (up to iron) that has enabled a star to remain in equilibrium and keep the crushing effects of gravity at bay. The only way that iron can be converted into another element is by adding energy to it, and

Fig. 1.22 The onion-skin model of the interior of a pre-supernova massive star. It is the growth of the iron (Fe) core that ultimately results in a catastrophic stellar death

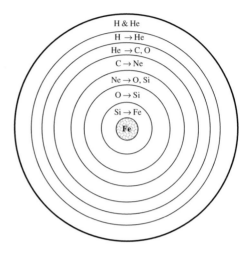

H & He

H → He

He → C, O

C → Ne

Ne → O, Si

O → Si

Si → Fe

Fe

this of course is that last thing that a star wants. The central temperature of the star is now several billion Kelvin, and the density is 5 million times greater than that of lead. Under these conditions copious amounts of neutrinos are generated, and these ghost-like particles quickly exit the star, traversing its interior within a matter of seconds, carrying away precious energy from the central regions.

Starved of fusion energy the core can do but one thing – contract under gravity. The contraction is rapid, taking just a fraction of a second to reach a temperature in excess of 10 billion Kelvin. At this point high-energy gamma-ray photons begin to interact with the iron atoms, leading to a process known as photodisintegration. In the blink of an eye, the iron nucleus is shattered into numerous helium atoms and neutrons, the core collapses completely, and the star is doomed. The loss of central pressure support ripples throughout the star, and the star literally begins to tear itself apart. A blast wave generated at the center propagates outward, and a further round of advanced fusion reactions can proceed to generate all the atoms up to the 94th letter of the matter alphabet (plutonium).

Many other isotopes and radioactive atoms are also produced at this stage, but these are not directly counted as distinct members of the Periodic Table of stable elements. The matter alphabet inventory is now complete, and a new generation of atoms is rushed into the interstellar medium.

Eventually, slowly but inevitably, the composition of the interstellar medium is enhanced in elements other than its original hydrogen and helium. Astronomers use the term heavy elements to describe all those elements other than hydrogen and helium, and each time a supernova explodes the interstellar medium becomes richer in heavy elements. This enhancement in elements beyond hydrogen and helium, of course, is crucial to our very existence. We live on a planet composed of silicate rocks, and we exist according to a chemistry centered on interactions between carbon atoms. We are very much children of the stars.

There is a wonderful cosmic recycling of material throughout the interstellar medium. Stars form out of the interstellar gas, go through their life cycle of fusion reactions, and – provided their initial mass is greater than about eight times that of the Sun – die as supernovae, placing heavy elements back into the surrounding interstellar gas out of which a new generation of stars can form, and so on. As cosmic time ticks by, so the interstellar medium is gradually depleted of hydrogen and helium, enhanced in heavy elements and reduced in overall mass. The mass of the interstellar medium is steadily reduced, since even during the supernova production stage a remnant neutron star core can survive. (We will have more to say about neutron stars in Chapter 7.)

Very massive stars might also leave remnant black holes upon going supernova. In addition those stars less than about eight times the mass of the Sun leave behind white dwarf remnants. All of these stellar end stages – white dwarfs, neutron stars, and black holes – lock away material from the interstellar medium, and there is accordingly a reduced amount of material to fuel the ongoing process of star formation (Fig. 1.23). The fate of the universe is to eventually become totally dark, but this won't occur until an unimaginably long time in the very distant future (as we shall see in Chapter 4).

Because of the increasingly harsh density and temperature conditions required to produce the atoms containing larger and larger numbers of protons within their nuclei, the abundance of such elements is much smaller than those atoms having fewer protons in their nucleus. According to mass fraction, the Sun, for example, is composed of 0.772 hydrogen and 0.209 helium, with the remaining 0.019 mass

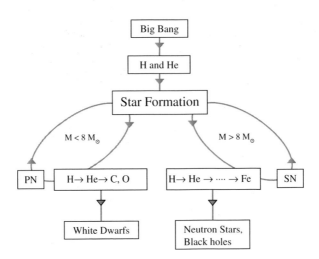

Fig. 1.23 The recycling of matter through star formation and stellar evolution. The recycling process represents the slow but steady conversion of the original hydrogen and helium into carbon and all other elements beyond. Those stars more massive than 8 $M_\odot$ undergo supernova (SN) disruption, while those less massive than 8 $M_\odot$ undergo a more gentle planetary nebula (PN) ejection phase at the end of their lives

fraction accounting for everything else in the matter alphabet. The mass fractions of carbon, nitrogen, and oxygen are 0.0039, 0.00094, and 0.0086, respectively, making oxygen the third most abundant element within the Sun. The Sun's mass fraction of uranium-238 (letter 92 in the matter alphabet) is estimated to be 1.4×10^{-10}, which amounts to about 3×10^{20} kg of material. It has been estimated that there are of order 3 billion kilograms of extractable uranium reserves on Earth, and if all of our electrical energy were generated through nuclear fission reactors, then the world's reserves would last for about 50 years at the present production rate. If one could extract the uranium from the Sun, and there is unfortunately no known way of doing this, then it would provide humanity with a near limitless fuel, its reserves being able to provide for an impressive 5×10^{11} years worth of energy generation at our present production rate.

The Hubble Deep Field

Astronomy has historically been concerned with the study of matter within the universe – or, more specifically, with matter that can either emit or absorb electromagnetic radiation. Visible matter, however, as we shall see in Chapter 5, forms only a small fraction of the universe's total mass inventory. In 1939 the pioneering astrophysicist Arthur Eddington made the wonderful statement, "I believe that there are 15747724136275002577605653961181555468044717914527116709366231425076185631031296 protons in the universe and the exact same number of electrons." This 80-digit number, which Eddington often wrote out in full, is actually derived from the product 136×2^{256}.

Perhaps the most remarkable feature of this number to the twenty-first century reader is that Eddington actually figured it out long-hand during an ocean liner crossing of the Atlantic – and got it right! This sumptuously large number, appropriately known as the Eddington number, was derived according to physical ideas that are no longer generally held to be viable, but we can with the aid of the modern-day Hubble Space Telescope and in the spirit of Eddington, make an attempt to calculate (admittedly with many assumptions) how many protons and electrons the observable universe might actually contain.

Figure 1.24 shows what has rightly become one of the most famous photographs within the Hubble Space Telescope gallery of images. Virtually every object in the picture is a galaxy, and the analysis of the data indicates that some 3,000 galaxy portraits, out to distances of order 1 billion parsecs, have been captured within the overall field of view.

The image in Fig. 1.24 covers an area of the sky equivalent to that of a square with sides 1/30 of a degree long – which is about 1/15 the diameter of the full Moon as seen from Earth. To cover the entire sky at this image scale the Hubble Space Telescope would have to produce a total of some 37 million images. Given that we would expect each of these deep field images to contain of order 3,000 galaxies (the reasons for thinking this will be discussed in Chapter 4), we can estimate that there are some 100 billion galaxies in the observable universe. In Chapter 5 we shall

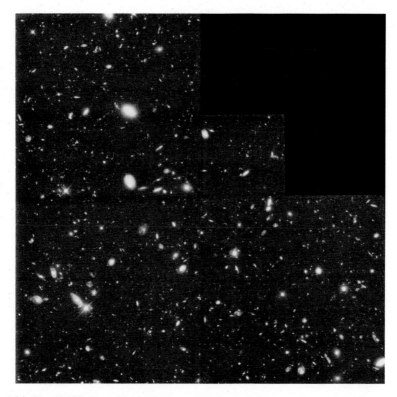

Fig. 1.24 The Hubble Deep Field. There are less than ten foreground stars in this image, and virtually everything else is a galaxy. (Image courtesy of NASA)

have much more to say about galaxies and their various characteristics, but for the moment let us assume that our Milky Way Galaxy is typical (some galaxies will be much bigger and some will be much smaller). It is generally taken that there are some 300 billion stars within our galaxy, so the number of stars in the observable universe will be of order 3×10^{22} stars (this number is simply our galaxy star count times the number of galaxies in the observable universe).

Given that our Sun (a pseudo-typical star) contains approximately 10^{57} protons (this is the Sun's mass divided by the mass of the proton), we can deduce that the stars within the visible universe contain approximately 3×10^{79} protons, and for charge neutrality the universe must contain the same number of electrons. Remarkably, although we have used both rounded and conservative numbers in our estimate, the number of protons in the observable universe, it turns out, is actually comparable to the Eddington number to order of magnitude.

The ancient Greek philosopher Archimedes would probably have appreciated our Eddington's estimate for the number of protons in the universe, not only for the fact that we can actually write the number down but also for its near infinite incomprehensibility. Indeed, working circa 220 B.C., Archimedes estimated in his famous text The Sand Reckoner that the universe was about 10^{14} stadia across (a diameter of about 2 light years – an uncomfortably claustrophobic universe by modern standards) and that it would require no more than 10^{63} sand grains to fill its entire volume. Let us, for sheer fun, take this calculation even further.

There are about 10^{20} atoms in a pure silica sand grain with each silica atom (which is expressed by the matter-word SiO_2) containing 30 protons. Accordingly, the sand-filled universe of Archimedes would contain of order 3×10^{84} protons. Although he didn't specifically advocate the filling of his universe with sand, Archimedes, it turns out, would have been disappointed even if granted the omnipotent power to try. Indeed, it would require some 100,000 universes, such as the one we see around us today, to generate the protons required to produce the sand necessary to fill the very much smaller cosmos envisioned by Archimedes.

Moving Forwards

In this chapter we have plumbed the depths of the cosmos, traversing the distance from the smallest quarks and electrons to that of the observable universe. We have traced connections and origins and explored some of the more remarkable numerical relationships that join the microcosm to the macrocosm – threads and lines of reasoning that even Apelles of Cos would have been impressed by. There is much that is known, and there are parts of the universe that we might even say we understand, but there are many more questions that have yet to be asked, and much more detail that has yet to be understood.

The remainder of this book will take us to the very edge of the known universe and, on some occasions, even beyond it. The next chapter will present an outline of the history of CERN and describe the build-up to the commissioning and construction of the LHC. Chapter 3 will present an outline of the standard model of particle physics – looking not only at the known but the unknown and speculative. Again and again, however, we will be drawn back to the mantra that within the macrocosm is a reflection of the microcosm, for the workings of the atom are writ bold and large across the great vault of the heavens.

Chapter 2
The World's Most Complicated Machine

For over 20 years now, if you listened very carefully, the ground below the verdant fields of the *Pays de Gex* region of France has trembled very slightly and perhaps, just perhaps, faintly hummed. Centuries ago vast armies trampled and rumbled across these same fields (Fig. 2.1), but today, thankfully, there is peace, and it is now a vast, many-nationed scientific army that works in the scenic vale of the Jura Mountains. The headquarters of this vast army of researchers, engineers, and fabricators is CERN, a complex hive of buildings, storage areas, heavy-lifting equipment, and numerous offices and workshops. It is Europe's exultant shrine to nuclear physics.

Below the ground, indeed some 100 m below the topsoil, there has been a great hurly-burly of activity these past years (Fig. 2.2). The Large Hadron Collider (LHC) has been growing, piece by piece, kilometer by kilometer, quadrant by quadrant, extending around the nape of the 27-km tunnel left vacant by the closure and dismantlement of CERN's last great experiment, the Large Electron–Positron (LEP) collider in 2000. Thousands of technicians, engineers, and research scientists have crawled, walked, and bicycled around the LEP, now LHC, tunnel, and for the last 20 years other researchers have toiled within university laboratories, offices, and meeting halls to perfect the equipment that constitutes the many faces of the LHC – the bright phoenix, full of promise, that is now arisen in the tunnel of the old LEP collider.

The story of the LHC began some 20 years ago, for even as the LEP collider was being built and commissioned, so its predecessor was being designed. Likewise, even before the LHC had produced its first experimental results, its predecessor – the next great collider – was being developed (see Chapter 7). Construction of the LHC proper began in the late 1990s, but after the inevitable delays and start-up rollbacks the experiment was finally readied for commissioning in 2008. As the grand start-up day approached, CERN garnered the attention of the world's media; the public was captivated and the physicists held their collective breaths.

M. Beech, *The Large Hadron Collider*, DOI 10.1007/978-1-4419-5668-2_2,
© Springer Science+Business Media, LLC 2010

Fig. 2.1 Map of LHC ring and surrounding area. The CERN headquarters is located at point 1, where the massive ATLAS experiment is also located. The *dashed line* snaking from the *lower left* of the image through to the *upper right* is the border between France and Switzerland. (Image courtesy of CERN)

The End of the Beginning

The CERN control room is packed. There are people everywhere. Some are sitting, some are kneeling, almost in a state of prayer, some are staring myopically at computer screens – deep concentration written across their brows. Others stand, or pace, casting their eyes intermittently towards the various electronic monitors indicating the progress of the initial beam. The minutes tick by. Checks and calculations, followed by more calculations and checks, are being made. Hands nervously fumble and turn cell phones and ipods. Some, especially the younger researchers, are dressed casually in jeans and T-shirts; others, the elder statesmen of science and the bureaucrats, pace the floor in well tailored suits sporting extravagant ties. Everyone, each in their own way, is displaying an air of excitement and tension as the cameras flash and journalists scramble for their interviews and sound bites. There is a steady hum of discussion, with the occasional voice briefly dominating the general background. All shapes, sizes, and ages of humanity are present, for today, September

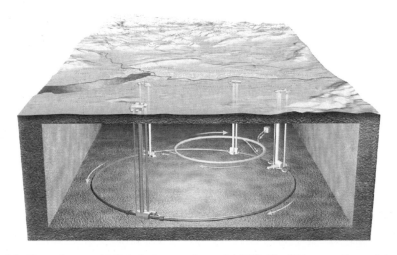

Fig. 2.2 The underground labyrinth that constitutes the LHC. The 27-km tunnel containing the proton beam steering magnets is angled at about 1.4% to the horizontal, resulting in the tunnel depth varying from 50 to 175 m below ground level. The smaller ring is the SPS accelerator that provides proton bunches to the LHC. The CMS experiment and access tunnel located near the small town of Versonnex is shown in the image foreground. Lake Geneva is shown on the left hand side of the image. (Image courtesy of CERN)

10, 2008, is first-beam day. After 20 years of hard preparatory work, history is about to be made.

In a machine as complex as the LHC there is no single start-up switch, and in reality it is millions of lines of computer code that control the many thousands of systems that must all function harmoniously and instantaneously if all is to go well. There can be no mistakes, and each component within the multiply linked equipment train must function to specification.

By 9 a.m. local time the CERN control room is abuzz with excitement. By 9:15 the tension is palpable. The start-up procedure is going according to plan, and Lyn Evans, LHC Project Leader, prepares (after one final press interview) to forward the beam from the Super Proton Synchrotron (SPS) (described below) storage ring to the second octant of the LHC (Fig. 2.3). An expectant hush descends. The countdown begins. A voice can be heard, "... three, two, one, ... Zero .... nothing?.... YES! ... yes ...". The time is 9:34. Amid a rapturous outburst of cheers and clapping a transient flash bursts onto a monitoring grid that is being displayed on a computer screen (Fig. 2.4). It is a dot, the fading echo of which indicates that the LHC has received its first packet of protons.

The first parcels of protons will be moved around the LHC octant by octant (Fig. 2.3) – initial little steps before the first giant leap. The steering of the beam will take the protons through the various channels of collimating super-cooled magnets (described below), where they will be "dumped" prior to a new set of protons being sent though the system and on into the following sectors. By 9:38 the proton beam will have progressed to a distance of 6.7 km around the 27-km loop of the

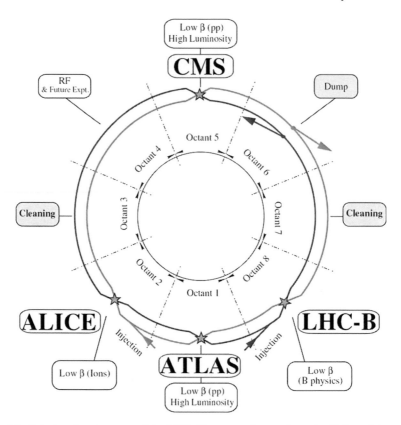

Fig. 2.3 Octant and sector map of the LHC. The proton beams from the SPS are injected at octants 2 and 8 close to the ALICE and LHCb experiment locations. A chain of many thousands of superconducting magnets steers two counter-directed proton beams around the 27-km LHC tunnel. (Image courtesy of CERN)

LHC. Time passes. At 10:00 the proton bunches are penetrating to the halfway point around the tunnel. The start-up is proceeding at a much faster rate than anyone had hoped for or even expected – all, it would appear, is going extremely well.

As the proton beam successfully completes the journey from one octant to the next and through one detector after another (Fig. 2.5), the excitement once again begins to grow. During an earlier interview LHC Project Director Lyn Evans had commented that he did not know how long it would take to circulate the first beam around the entire 27-km length of the LHC. "It took us 12 h to circulate a beam around the Large Electron Collider," (LEP) he noted, and no predictions were being offered for the time that it might take to ensure that all the LHC collimators were working synchronously. Bottles of champagne had already been opened after the first protons entered the LHC, but the real triumph of the day was realized at 10:28 local time, a mere 50 min after the first proton packet had been injected; this was the time when the first beam of protons completed the inaugural clockwise journey

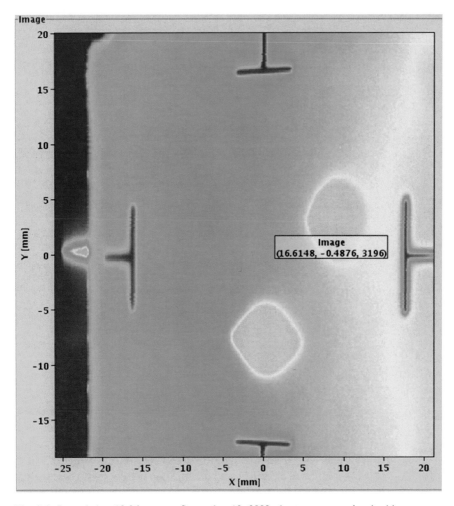

Fig. 2.4 Recorded at 10:26 a.m. on September 10, 2008, the two *orange dots* in this computer screen image signify that the first proton beam has completed a full turn around the LHC storage ring. (Image courtesy of CERN)

around the entire 27-km loop of the LHC. CERN Director General Robert Aymar summed up the first beam success in grand Churchillian terms; "C'est la fin de la debut" – *This is the end of the beginning*. By 10:30 a proton beam had completed three full turns around the LHC. The leviathan was coming to life, and hopes for the future were riding high. CERN had caught the world's attention and the highly publicized and much media-hyped first-beam day was proceeding just marvelously, thank you very much.

Having circulated the proton beam around the LHC in a clockwise direction, the next step was to send a beam around in the opposite, counterclockwise direction.

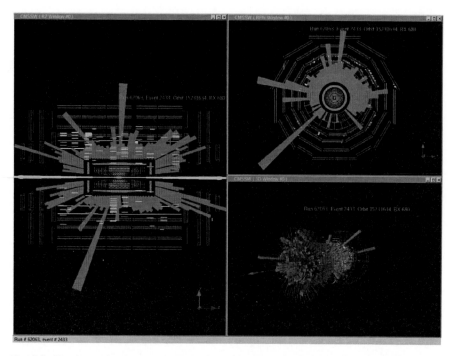

Fig. 2.5 First-beam day images from the CMS showing a spray of particles detected in the system's calorimeters and muon chambers after the proton beam was steered into a terminating tungsten block. (Image courtesy of CERN)

No beam collisions were going to take place, but the successful completion of such a feat on first-beam day would be a great publicity coup. On cue, at 3 p.m. local time, the good news continued at CERN, and the first counterclockwise beam completed its first successful circumnavigation of the 27-km LHC ring. First-beam day in Geneva had in the vein of Revelations ii. 17, turned into a white stone day.

Disappointment and Setback

When the world's media is watching there is nowhere to hide if things go wrong. It was perhaps a risky gamble in the first place to announce an official first-beam day. A machine as complex and as innovative in new technologies as the LHC is hardly likely to work perfectly when first turned on. There was no denying, however, the frustration and disappointment that followed in the wake of the first-beam success.

The first major equipment failure was recorded on the night of September 12, when a 30 metric ton power transformer malfunctioned, causing some of the magnets to warm above their optimum working temperature. This failure, however, was a relatively minor and straightforward problem to fix, and within a matter of days

Fig. 2.6 A view of sector 3-4 in the LHC tunnel where the September 19, 2008, malfunction occurred. (Image courtesy of CERN)

the entire transformer had been replaced. The next system failure was much more problematic.

At noon on September 19 there was what the CERN press office laconically called an "incident in LHC sector 3-4." The incident involved the melting of an electrical coupling between two steering magnets and the triggering of a massive magnet quenching effect that resulted in some 1,000 kg of liquid helium being vented into the LHC tunnel (Fig. 2.6). In essence the coupling fault caused the magnets to warm up and to abruptly lose their superconducting state. CERN Director General Robert Aymar described the failure as "a psychological blow," since in order to fix the problem the magnets in the affected sector would need to be brought back to ambient tunnel temperature and then re-cooled once repaired – a process that would take several months. Not only was the time delay a blow to morale, what was perhaps more worrying was the realization that the problem might not be an isolated one. Indeed, it soon became apparent that a full system investigation was needed, and with the obligatory winter closure for maintenance rapidly approaching it was decided on September 23 to call a halt to any further LHC commissioning. After a 9-day inaugural run the LHC was shut down for repairs.

Repairs and upgrades to the magnetic steering system began immediately upon closure, but it was going to be a long delay before the LHC might see action again. At a workshop held in Chamonix, France, in February 2009, CERN managers decided upon a restart schedule that would see the LHC begin operations again towards the end of September 2009, with full power collisions not being scheduled to take place until at least 2011. The new schedule also called for the LHC to run through the winter months of 2010, a time earlier scheduled for routine maintenance and repairs. This decision alone would be expensive, since the cost of energy is much higher during the winter, and at least 7.5 million Euros (about $10 million) would be added to CERN's electricity bill. But, as Steve Myers, the director of accelerators and technology at CERN mindfully observed, "We built this machine to operate it – if you buy a Rolls-Royce, you can afford to put the petrol in." Indeed, the cost of the electricity during the winter extension is small-fry compared to the LHC's estimated $6.5 billion price tag.

As the summer of 2009 progressed so, too, did the work on the LHC upgrades. The last of the re-vamped steering magnets was lowered back into place by the end of April, but numerous smaller problems and delays slowed the re-commissioning process. A new start-up date was set for November 2009, and, as history will record, at 13:22 GMT on November 23 the very first collisions were detected in the ATLAS experimental chamber.

The initial November re-start collisions were performed with the beam energy held back to a mere 450 GeV, but with the system commissioning running so smoothly it was soon decided to ramp up the power, and just before midnight on November 29, 2009, a world-record breaking collision energy of 1.18 TeV was obtained. The LHC accordingly became the most powerful particle collider ever built. As the labyrinthine system continues to prove its reliability so the plan is to increase the energy within its beams throughout 2010 to a maximum collisional energy of 7 TeV. According to present projections it will likely be at least mid-2011 before full 14 TeV collisions will take place. In spite of the relatively low collision energies employed in the re-start program, breaking what is probably a speed record for the dissemination of experimental information, the first research paper based upon data collected with the LHC was accepted for publication on December 1, 2009. The paper was submitted by a consortium of several hundred researchers, all working on the ALICE experiment, to the *European Journal of Physics*, and contains an account of the measured density of charged particles within the detector – a humble paper by particle physics standards, but the first results from an ever-improving LHC.

Now that the LHC experiments are actively gathering data the new discoveries of physics awaits, and the hopes of the researchers are once again riding high. Commenting to *New Scientist* magazine in its August 7, 2009, issue, Greg Landsberg (Brown University in Providence, R.I.) observed in optimistic tones that, "Nature is full of surprises, and something exciting and possibly unexpected could happen at 7 TeV. Extra dimensions [to be described in Chapter 7] could easily open up at that energy." Good things come to those that wait, and the great leviathan that is the LHC is now awake, roaring and full of promise for the future.

Court Case Number 1:2008cv00136

Given that particle colliders are built and indeed funded on the very basis that they will take physics into unknown areas of study, the question of safety is a natural one to ask. How do we know that the LHC won't produce some bizarre Earth-destroying particle or trigger an event that will result in our collective doom? Indeed, miniature black holes, strangelets, and magnetic monopoles (to be described below and in Chapter 7) might all be produced by the LHC – and the appearance of such exotic objects, some researchers have suggested, could be disastrous. Should all humanity, therefore, be in fear of an existential catastrophe? And, should we really be calling for the immediate closure of the world's greatest physics experiment now that it is successfully running?

There has been much media discussion about the possibility of an LHC-induced catastrophe occurring. The topic is indeed an eye-catching one, making for good press, and it brings out many complex issues concerning science and society. This being said, emotions and hyperbolae have been running high. None of the recent furor surrounding the commissioning of the LHC should surprise us, however. Here, for example, is how the March 9, 1931, issue of *Time* magazine opened a story concerning the first atomic accelerator experiments that were then beginning to come online: "Predicting the end of the world has been immemorially the privilege and pastime of Religious fanatics and charlatans. In modern times such predictions have been the province of loose-spoken scientists and the sensational Press. The cry of modern world-enders is that if anyone ever succeeds in exploding one atom of matter, the whole universe will go off like a bunch of firecrackers." The tone of the passage is perhaps overly dismissive and somewhat derogatory, but it exemplifies the level of passions that can run in such science and society debates. Within a year of the *Times* article being published, however, the Cavendish Laboratory-based research team of Ernst Walton, John Cockcroft, and Ernst Rutherford successfully "split the atom" under controlled conditions for the very first time (on April 14, 1932), and the universe, in spite of the dire warnings of the "world-enders," continued to exist. Fast forward 77 years.

Filed on March 21, 2008, before the Hawaii District Court, Judge Helen Gilmore presiding, court case 1:2008cv00136 brought together plaintiffs Luis Sancho and Walter L. Wagner and defendants the US Department of Energy, Fermilab, CERN, and the National Science Foundation. The demand of the plaintiffs was that the LHC should not be set into operation before it had been proven to be absolutely safe. We will deal with the specific safety concerns in a moment, but suffice it to say, the plaintiffs were not especially worried about the health and safety of the CERN engineers and researchers. Rather, and in some highly laudable sense, the plaintiffs were concerned about the potential death of all of humanity, an Earth-crushing 6.8 billion people, as well as the destruction of Earth itself. It is literally an existential threat, and the ultimate demise of humanity, that the plaintiffs were worried about.

At least two fundamental issues were raised by court case 1:2008cv00136. First, can one prove that any experiment is absolutely safe, and second, who should decide whether an experiment is safe? Although such questions are important and fundamental, there are, once all the bickering is said and done, only two practical answers that can be offered. Firstly, it is impossible to prove beyond any reasonable doubt that any experiment is absolutely safe; indeed, upon reasoned reflection, the question itself is not well founded, and second, if anyone is capable of deciding upon the safety issues of an experiment, then it must surely be the people who designed it and who fully appreciate the high-energy physics that it has been designed to test.

There are, accordingly, very few people alive today who are actually well versed in the field of collider and particle physics. As a pure guess, let us assume that there are 20,000 such experts worldwide. This guess is probably an overestimate, and the real number of actual particle physicists is probably going to be much smaller. In other words, less than 0.0003% of the world's population is likely to have a detailed working knowledge of how the LHC is constructed and what the likely consequences of running the actual experiments are going to be.

It is for this reason that the argument of the plaintiffs in court case 1:2008cv00136 falls well short of being reasonable. At the core of the problem is a very common human failing – a poor ability to understand and appreciate risk. Before seeing why the LHC offers absolutely no existential risk, we should briefly mention some of the specific problems that concern the plaintiffs. They are: (1) the production of strangelets that might catastrophically convert all normal matter into a new lower-energy form, (2) the production of miniature black holes that might literally devour the entire Earth, and (3) the generation of magnetic monopoles that might trigger the destruction of protons and the collapse of otherwise stable matter. If the theories that some researchers have published and popularized relating to these specific phenomena are correct, and they do actually pan out, then it is fair to say that we are in for some serious trouble. But are we really at serious risk?

If there is even the smallest probability that the destruction of the entire Earth might come about through the operation of the LHC, then why has its construction and commissioning been allowed? The answer, in fact, is quite straightforward. Our safety is assured by the fact that the world is already 4.5 billion years old – indeed, the very existence of the Sun, Moon, planets, and stars are glittering testaments to the safety of the LHC. Nature, as we shall discuss in greater detail in Chapter 7, has already run nearly half a million experiments, similar to that which will be conducted at the LHC during the next decade, in our upper atmosphere, and Earth, philosophical quibbling aside, is still here and we are still very much alive.

The agents of nature's collider experiments are cosmic rays. These small charged particles pervade the entire galaxy, and while diminutive in mass they move through space at close to the speed of light. Indeed, the energy carried by some cosmic rays can be colossal and far exceed that which will be realized by the LHC when operating at full power. The Moon's surface and Earth's atmosphere have been continuously strafed by high energy cosmic rays ever since they first formed, and yet not one of the cosmic ray encounters has resulted in our destruction through the generation of strangelets, miniature black holes or magnetic monopoles. Such exotic objects may well have been produced by cosmic ray collisions, but they have not resulted in any catastrophic destruction.

Afterwards

On October 1, 2008, Judge Helen Gillmor dismissed court case 1:2008cv00136, arguing that the Honolulu court lacked any jurisdiction over the European-based CERN accelerator. Interestingly, however, Judge Gillmor commented that the suit revolved around a "complex debate" and that the issue was of concern to the whole world. Perhaps no one would disagree with these sentiments, but it seems a little ironic that the case was thrown out, not because it had no merit but because of a legal issue relating to the jurisdiction domain of the court. The message from the Honolulu court appears to be that the world might well end when the LHC is turned on, but trying to prevent it from ending isn't their specific problem. Well, was all the fuss over this special court case for nothing? The answer is probably no; it is certainly right to question what results the LHC experiments might potentially produce, especially if they might result in an uncontrollable disaster, but the conclusions derived by several detailed safety review committees are clear – there is no existential threat posed by running the LHC.

With all of the above being said, the LHC is not out of the legal jungle just yet. Indeed, another lawsuit was filed against CERN on November 20, 2009, by a group calling itself "conCERNed International." Representing several different organizations and individuals conCERNed International has filed a complaint with the Human Rights Commission of the United Nations arguing specifically that "the CERN member states, especially Switzerland, France and Germany, [have not] carried out their legal responsibilities to ensure citizens' safety." The new complaint calls for the establishment of a CERN-independent safety review committee and also calls into question the legitimacy of the cosmic ray analogy (described more fully in Chapter 7) used by previous safety review committees to argue that the LHC is perfectly safe. The debate continues, but it is now purely a legal one, since the LHC is already in operation.

Overview: A Proton's Journey

The title of this chapter is no idle exaggeration; the LHC is the most complicated machine ever built by humanity. It dwarfs by comparison anything else that has ever been constructed. Everything, including its size, technical complexity, scientific innovation, and exactitude of construction is on the grandest of scales. We are honored to live in an epoch when such a machine, if that really is the right name for the LHC, can be built. But enough eulogizing already – the author's thoughts on the LHC are presumably clear by now.

For all its incredible innovation the LHC may be thought of as the superposition of four basic units: the accelerator system; the proton beam confinement and focusing system; the particle collision detectors; and the data processing and monitoring system. All of these systems are massive and complex in their own way, and all must function in near absolute synchronicity. Indeed, the LHC works on the hyper-accelerated timescale of atomic interactions, where many thousands of things can happen in just a few hundredths of a nanosecond (1 ns equals one billionth of a second – see Appendix 1), and during the merest blink of an eye millions of particle creation and decay events will have taken place.

Before describing the function and design of the various LHC building blocks, let us go on a metaphorical package holiday tour of CERN's newest collider. Our whistle-stop journey will follow the path of the protons as they wind their way around the various accelerator rings towards an ultimate collision event and destruction in a spray of elementary particles in one of the experimental chambers. The journey will also be a passage through time, since the LHC is fed protons by its historical predecessors.

The proton's journey to the LHC begins in a small red cylinder containing hydrogen gas (Fig. 2.7), located inside of a rather non-descript cabinet next to the LINEAC 2. LINEAC 2 is a linear accelerator whose job is primarily to strip the hydrogen

Fig. 2.7 The hydrogen supply cylinder – the starting point of the LHC experiment. At full beam loading about 2 ng (2×10^{-12} kg) of hydrogen will be consumed by the LHC per day. (Image courtesy of CERN)

Fig. 2.8 The many twists and turns of the proton beam supply route for the LHC. The protons begin their journey at point P in the diagram, looping through the Proton Synchrotron and the Super Proton Synchrotron before being fed into the LHC confinement ring. (Image courtesy of CERN)

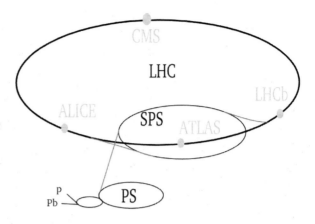

atoms (fed in from the gas cylinder) of their electrons and to accelerate the resultant protons along a straight (hence linear) track, to about a third the speed of light before passing them into the Proton Synchrotron Booster (PSB).

Following further acceleration in the PSB the protons are fed into the Proton Synchrotron (PS) ring (Fig. 2.8). Built in the late 1950s the PS ring is, by modern standards, a paltry 200-m across. Its role in the LHC, however, is to continue adding more and more energy to the protons. When they leave the PS and are injected into the Super Proton Synchrotron (SPS), the protons have acquired a speed of nearly 9/10 that of the speed of light. The SPS was completed in 1976 and has an accelerating ring system that is about 2.25 km across. It is from the SPS ring that the LHC is fed. Once positioned within the LHCs 27-km long confinement system the protons will be accelerated to their top speed of 0.999999991 times that of light.

When fully operational, something like 2 million millionths of a kilogram (2 ng) of hydrogen gas will be consumed per day, and it will take about 10 min to fill the LHC ring. Rather than being a continuous stream of protons, however, the LHC beam is actually composed of bunches of protons.

At full capacity there will be 5,616 bunches spread around the 27-km track of the confinement ring – 2,808 proton bunches moving clockwise around the ring, and 2,808 bunches moving counterclockwise around the ring. Each bunch will be composed of about 100 billion protons, and the physical spacing between each bunch will be about 7.5 m. Given their near light speed it will take about 1/11,000 of a second for each proton bunch to circle the entire LHC ring system, and one proton bunch will enter and collide within each experiment hall every 25 ns. The two counter-rotating proton beams are brought to a focus at each detector node, and at these collision points the beam confinement region is reduced to a width smaller than that of a human hair – of order 40 million proton bunch crossings will occur at each experimental focal point per second. When at full operating power something like half a billion proton–proton collisions will occur in each detector every second, corresponding to about 23 collisions per bunch interaction. In any one 25 ns instant there will be something like 1,600 elementary particles in each LHC detector, and all of these particles need to be identified and accurately tracked.

There is probably not enough computing power in the entire world to record and analyze all of the collision events that will take place per operating day at the LHC. There are nearly 100 million data channels emanating from the LHC experiment halls, and if all the information streaming along these channels were to be saved, then the data would fill the equivalent of about 100,000 CDs per second. There is literally too much data to handle.

Luckily, the vast majority of collision events are of no specific interest to the LHC researchers (they relate to well understood physics rather than the new), so various triggering protocols have been developed to pick out the rarer and more unusual events. The filtering process reduces the near half-billion event count per second to a more manageable 100–200 events per second that need to be studied. For each of these one hundred or so special events per second there will be a spray of associated particles within the detector, as well as a background host of other particles associated with the other events that have no specific interest. Even with the triggering mechanism in place of order 250 MB of data will be gathered from the LHC experiments every second, amounting to about 10 petabytes (10^{16}) of data being archived per year. In terms of CD stack height the annual LHC data output would stretch 20 km into the sky – a height two times greater than that of Chomolungma.

To solve the colossal storage problem the data from the LHC will be archived on magnetic tapes (Fig. 2.9). In some sense, tape storage is akin to a blast from the ghost of technology past, but it is a cost effective and highly reliable medium upon which to store a very precious data cargo. The LHC computing grid is a formidable construct within its own right, and its job will be an intensive one since a highly complex reconstruction process must be applied to every collision event that is recorded. Indeed, rather than attempt to analyze all the experiment data at CERN, much of it will be sent to other institutions around the world for final interpretation.

Computing of one form or another has always played an important role at CERN, and indeed, the now ubiquitous worldwide web saw its fledgling origins there. The idea for such a server and open Internet access system was first proposed by British physicist Tim Berners-Lee in a short information memo written in March of 1989. Following further development of the idea with Belgian computer specialist Robert Cailliau, the web became reality in late 1990. The world's first-ever web server can still be accessed under the URL (uniform resource locator, to those in the know) by initiating the hypertext transfer protocol (http) command: http://info.cern.ch. The rest, as the saying goes, is history, and by July of 2008 the Google search engine had indexed over one trillion unique web page addresses worldwide.

The web is actually being exploited by researchers at CERN to enable a volunteer computing network to perform simulation calculations. Called LHC@home the idea is to utilize the idle time of home computers to carry out numerically intensive calculations that don't require the transfer of large data files. The network was initially set up in 2004, and the first user application was a program called *SixTrack,* which simulated the stability of particles orbiting around the LHC storage ring. Additional

Fig. 2.9 A small portion of the vast, robot-fed magnetic tape storage area that will archive the LHC's "events of interest" data set. (Image courtesy of CERN)

information about the LHC@home network and sign up details can be found at http://lhcathome.cern.ca/.

Of order 120 MW of power will be required to run the fully operating LHC, and this amounts to some 800,000 MW h of energy being consumed per operational year (typically lasting 270 days). The estimated annual electricity bill will come in at about 19 million Euros (some $25 million). Most of the energy consumed by the LHC will go into powering the superconducting magnets that are responsible for steering the proton beams around the 27-km storage ring. When fully loaded with proton bunches the total amount of energy contained within the circulating beam of particles will be 362 MJ; this is equivalent to the kinetic energy of a 400-ton train traveling at 153 km/h, or, the amount of energy required to melt an 88-cm sided cube of water-ice. No matter which way one tries to look at such a quantity of energy it is very large, and accordingly a great deal of attention has been directed towards avoiding catastrophic heating events from occurring within the magnets. Indeed, a machine protection system has been developed to continuously monitor all of the critical components and to initiate a beam-abort procedure if a system failure appears imminent.

The Journey to the LHC

Isaac Newton once famously quipped that if he had seen farther than others it was because he had been able to stand upon the shoulders of giants. For Newton this was an uncharacteristic nod to the fact that no scientist really works in isolation, and great innovations typically build upon the ideas only hinted at by others. In the case of CERN, however, the LHC truly is a towering triumph because it is built upon and around the shoulders of many brilliant and dedicated researchers – the giants of the experimental physics world.

That CERN came to exist at all is a remarkable story in its own right, and its origins stretch back to the innocent optimism that pervaded post-World War II Europe. Physicist Louis de Broglie, who we encountered in Chapter 1, in fact, spearheaded the first official proposal to the European Union to fund and establish an international atomic physics laboratory in December of 1949. Subsequent hard work and much debate resulted in the CERN Convention being signed by twelve founding member states in July of 1953. Geneva was chosen as the site for the laboratory, and the first construction at Meyrin (see Fig. 2.1) began on May 17, 1954 (Fig. 2.10).

Fig. 2.10 Construction begins at the Meyrin, Switzerland, site on May 17, 1954. (Image courtesy of CERN)

One of the first large-scale projects to be developed at CERN was the 629-m diameter Proton Synchrotron, which accelerated its first protons on November 24, 1959. The PS became the heart of the particle physics program at CERN, and it is still operational to this very day, forming part of the proton feed system to the LHC (Fig. 2.8). The 7-km diameter Super Proton Synchrotron was commissioned at CERN in 1976 and was the first accelerator ring to cross the Franco-Swiss border. It was with the SPS that CERN researchers in 1983 were able to discover the W and Z bosons that mitigate the weak nuclear force (described in Chapter 3). Tunnel boring for the 27-km circumference Large Electron Positron (LEP) collider began in 1985, with the entire system being commissioned for scientific use in July of 1989. Prior to the construction of the 50.5-km long Channel Tunnel (completed in 1994), the LEP ring-cavity was the largest civil engineering project ever undertaken in Europe. The LEP initiative ran a highly successfully research program for 11 years but was finally closed in November 2000 to make way for the construction of the LHC.

Collider Basics

Why build, at considerable public expense, the LHC? This, of course, is a loaded question, and it invites many possible answers based, invariably, upon the respondent's particular take on the state of the world, the economy, and society. The question, however, is intended as a physics one.

What is it that makes the LHC better than all the other existing and previously built colliders? The answer is simple: energy. The more energy within the colliding beam of particles, the smaller the atomic scale that can be explored and the more massive the resultant elementary particles that might be produced.

Although the minds-eye image of the LHC experiment is one of protons colliding, it is really the quarks and gluons (described in Chapter 3) within the protons that are interacting, and the most hoped for particle to be discovered is that of the Higgs (also to be described more fully in next chapter), which has a minimum expected mass of about one hundred times that of the proton. To enable quark–quark interactions and to produce Higgs particles, the protons within the LHC beam need to be moving fast – indeed, almost as fast as light itself.

In the classical physics domain the energy of motion, the so-called kinetic energy (K), is related to the mass (m) and velocity (V) of the particle through the relation: $K = \frac{1}{2} mV^2$. For a given mass, therefore, the only way to increase the kinetic energy is to increase the speed; by doubling the speed of a particle the kinetic energy increases by a factor of $2^2 = 4$.

At first thought it might seem that there is a natural limit to how much kinetic energy a particle can have since, as Einstein argued in 1905, the fastest speed that any object can move is at the speed of light: $c = 299{,}792{,}458$ m/s. Although this speed limit is a true limit, it turns out that there is no specific upper limit to the amount of energy that a particle can have. Once again, the predictions of classical physics have to be modified. The reason for this seemingly strange result of having a limited speed but an unbounded energy is explained through Einstein's special theory of relativity. For a particle moving with a velocity V close to the speed of light the kinetic energy of motion is expressed as $K = (\gamma - 1) mc^2$, where $\gamma = 1/(1 - \beta^2)$ and $\beta = V/c$. It was upon deriving this relationship that Einstein discovered his famous result that an object of mass m has an associated rest mass energy of $E = mc^2$. The key point about the special relativistic version of the kinetic energy equation, however, is that as the velocity V approaches the speed of light so γ

becomes larger and larger and the energy term grows bigger and bigger. Technically, if the velocity actually equaled the speed of light then the energy would be infinitely large – and this, of course, cannot happen, since it would require an infinite amount of energy to accelerate the particle to the speed of light in the first place (Fig. 2.11).

Fig. 2.11 Energy versus velocity relationship for particles moving close to the speed of light. The energy is expressed in terms of the particle rest mass (mc^2), while the velocity is given as a fraction of the speed of light ($\beta = V/c$)

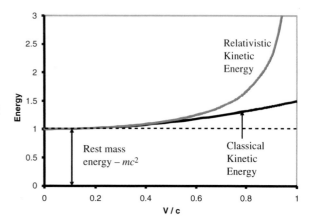

However, the point is, the energy carried by a particle becomes very large once the velocity approaches that of light, and in the LHC the protons within each beam will be accelerated to a top speed of 0.999999991c, corresponding to an energy of 7 Terra electron volts (7 TeV). Using Einstein's mass-energy equivalency formula, the mass of a proton is 938.2720 MeV, hence each proton within the LHC beam will have an associated energy about seven times larger than its rest mass. In addition to this, the great utility of smashing two oppositely rotating proton beams together is that the resultant energy from the collision is the sum of the beam energies, and in the case of the LHC this will be of order 14 TeV.

The all-important Higgs particle that LHC researchers will be searching for has an estimated mass in the range 100–200 GeV, and this explains why the proton beams have to carry so much energy. Only after the energy concentration during a collision exceeds the rest mass energy of the Higgs will its production be likely. Why the Higgs is actually important will be made clear in the next chapter, but for present-day particle physicists it is the Holy Grail – the highly sought after prize that will explain why all matter, you and me, the universe, has a corporal mass. The (to be described) Higgs field in some literal sense is the very muscle that animates the particle physics world.

Even when the energies are right, however, Higgs-producing events are going to be rare. Indeed, when operating at full power perhaps only one Higgs production event will occur in the LHC per half-day interval of time. It will take several years of near continuous operation before the LHC will have generated enough data to be statistically certain that the Higgs has truly been found.

From a physics perspective a collider needs to do essentially two things: constrain and steer the particles to their appropriate collision points in its associated experimental halls, and to accelerate the particles to high speed. By using charged particles (such as protons or electrons) both the steering and accelerating processes can be achieved with magnetic and electric fields, respectively (Fig. 2.12). The key physical processes being applied are illustrated in Fig. 2.13. The path of a charged particle is deflected according to the strength and polarity of the magnetic field through which it is passing – as illustrated in Fig. 2.13a. Likewise, the speed of a charged particle is increased when it encounters a region containing an electrical field – as illustrated in Fig. 2.13b.

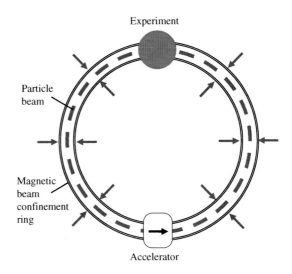

Fig. 2.12 The essential constituents of a particle collider system. Magnetic fields (shown schematically as *radial arrows*) are used to constrain the path of the particles, while an applied electric field (*heavy horizontal arrow*) is used to accelerate them to high speed. See also Fig. 2.13

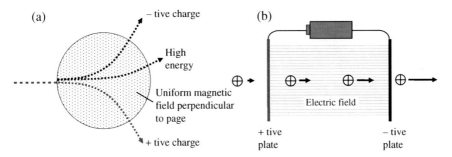

Fig. 2.13 (a) The effect of a magnetic field upon the motion of a charged particle. Positively charged particles will have paths curved in the opposite direction to negatively charged ones, and the greater the energy of the particle the smaller is its curvature. In this diagram the magnetic field is imagined to be coming out of the page towards the reader. (b) The particle accelerating effect of an electric field. The *arrow* length indicates the increasing speed of the particle

The energies associated with the proton beams that will be circulated within the LHC are so large that exceptionally strong magnetic fields are needed to constrain them. Indeed, the design of the magnetic collimators was one of the most complex challenges faced by the LHC engineers. When operating at full power the strength of the magnetic fields in the steering magnets will need to be of order 8 T – a field strength that is nearly 30,000 times greater than that of Earth's geomagnetic field. In addition to confining and steering two counter rotating proton beams, two oppositely orientated magnetic fields need to be maintained. To achieve all of these design requirements, and for the finished components to physically fit into the LEP tunnel, new and innovative superconducting magnets were developed.

Figure 2.14 shows a cross-section slice through one of the 1,232 dipole magnets that will help guide the proton beams around the LHC tunnel. Each magnet is composed of niobium-titanium coils that run along the inner core of the magnet (shown in red in Fig. 2.14 inside of the green non-magnetic collars). The magnetic coils are made of cable consisting of 36 twisted strands of 15-mm wire, with each wire being in turn composed of 6,400 individual filaments; each filament is about 1/10 the thickness of a human hair. Within the LHC magnetic steering system there are about 7,600 km of cable, composed of 270,000 km of strand.

LHC DIPOLE : STANDARD CROSS-SECTION

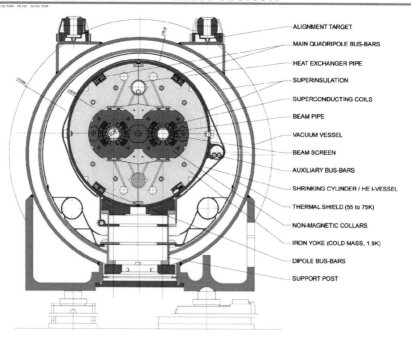

Fig. 2.14 Cross-sectional slice through one of the LHC's dipole steering magnets. Each magnet system is 15 m long, about 1 m in diameter, and weighs in at 25 tons. (Image courtesy of CERN)

The niobium-titanium coils become superconducting, meaning that they present zero resistance to any applied electric current and exhibit virtually no power loss, once their temperature drops below 10 K (a very crisp −263.2°C). The magnets at the LHC will operate, in fact, at the even cooler temperature of 1.9 K (−271.3°C), making them colder than the microwave background that pervades all of outer space (see Chapter 4). One of the greatest technical challenges associated with the construction of the steering magnets was that of keeping them cool. Indeed, the LHC magnetic steering ring constitutes the world's largest cryogenic system, with each magnet being refrigerated by super-cooled liquid helium. About 120 tons of helium is required to cool the entire magnetic confinement system, which has a total mass of about 37,000 tons, and it takes anywhere from 3 to 6 weeks to cool down each sector of the ring (recall Fig. 2.3) to its operational temperature.

It took over 4 years to assemble and install all the magnets in the LHC tunnel. The first magnet arrived at CERN in January 2003, but it wasn't until March of 2004 that half of the superconducting cable required to build all the magnets had even been manufactured. The last magnet, however, was lowered into place in May 2007, and it was July 2008 before the entire beam confinement system had been fully closed and reduced down to its hard-vacuum operating conditions (Fig. 2.15).

Fig. 2.15 Technicians test one small segment of the 27-km long, super-cooled, cryogenic magnetic proton beam confinement system. (Image courtesy of CERN)

The entire 27-km ring of magnets had been brought down to their superconducting, 1.9 K operating temperature by August 2008, ready and waiting for September 10 and first-beam day (as described earlier).

While the proton packets circulate around the LHC ring system, they will suffer some inevitable losses of energy and degradation. Even though the beam tube is under a hard vacuum there will always be some residual atoms with which a proton might interact. Likewise some protons will inevitably be perturbed out of the nominal beam orbit, crashing eventually into the beam line wall. In addition, since the protons carry a charge and are constrained to move along a circular path they will radiate synchrotron radiation, resulting in a loss of energy. To counteract these effects the LHC ring system contains two beam-cleaning regions, located in quadrants 3 and 7 (see Fig. 2.3). At these locations any stray protons are filtered out of the beam.

In order to boost the proton beam energy a set of cryogenic radiofrequency (RF) cavity systems was installed in quadrant 4. These chambers apply a particle accelerating electric field of 5 million volts per meter at a frequency of 500 MHz (the radio part of the electromagnetic spectrum). The RF cavities will accelerate the proton packets, after they have been injected from the SPS, to their final top speed prior to the commencement of collisions, and they will also shape the proton packets, making sure that the protons remain tightly grouped together; this is important with respect to maintaining as high a luminosity at the experiment collision points as possible. Once collisions commence at the various experiment nodes and data extraction begins the beam intensity will gradually fall, since protons are being actively lost through collisions. But it is estimated that an optimal run time, before a new set of proton packets has to be injected and accelerated, will be about 10 h.

The Detectors

There are four main detectors located around the LHC beam line as well as a number of smaller experiments that will run at intermittent times. Two of the major detectors, ALICE and LHCb, will be described in Chapter 4, while a smaller experimental station, LHCf, will be described in Chapter 7. In this chapter we will concentrate on the colossal ATLAS and CMS detectors as well as the more diminutive, but no less important TOTEM experiment.

The job of the superconducting magnets and the 27-km ring system is to prepare packets of counter rotating protons that can be brought to a known collision point within a detector. The job of the detectors is then to capture as much information as possible about the collision debris. The spray of particles produced in a collision must be closely tracked and identified, and the amount of energy that each particle carries must also be measured as accurately as possible. To achieve these ends researchers use magnets, once again, to distinguish between positively and negatively charged particles, and by carefully recording the path of the particles

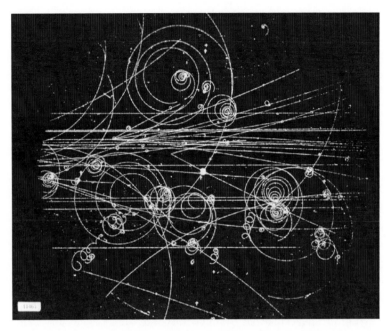

Fig. 2.16 The curvature of charged particle paths, when moving through a magnetic field, is illustrated in this bubble chamber image. Particles with low momentum produce *highly curved spiral tracks*; particles with higher momentum produce only *slightly curved tracks*. Positive and negatively charged particles curve in opposite directions to each other. (Image courtesy of CERN)

as they move through the detector's magnetic field their specific momentum can be determined; high momentum particles move along almost straight tracks, while low momentum ones make tight spiral turns (Fig. 2.16).

The energies associated with each of the collision-produced particles are measured with highly sensitive calorimeters. Since, however, different particles have different interactions and penetrative characteristics the calorimeters have to be carefully nested, one inside of the other. Figure 2.17 shows the basic design arrangement for a multi-layered calorimeter. The innermost calorimeter, located closest to the proton beam, determines the energy carried by electromagnetic radiation (photons) and electrons. The next layer measures the energies associated with hadrons, such as protons and neutrons, while the outermost calorimeter measures the energies associated with muons (which are short-lived leptons – see Chapter 3).

Each calorimeter is constructed from very specific detector material, and, of course, all have to work together in perfect synchronicity. From the very outset it was clear that the LHC detectors would need exceptionally good tracking capabilities as well as a high granularity so that the paths and energies associated with the decay particles could be unraveled, tagged, and ultimately traced to one specific collision event.

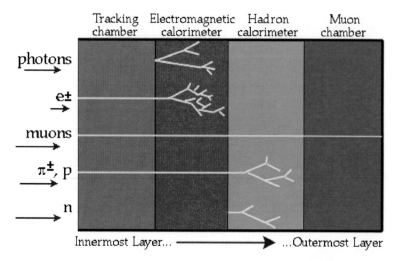

Fig. 2.17 The nested arrangement of calorimeters used to determine particle energies. (Image courtesy of CERN)

The layout of the ATLAS (A Toroidal LHC ApparatuS) detector is shown in Fig. 2.18. The detector is a behemoth, with the detectors and support frame weighing in at some 7,000 tons. Larger than most houses, ATLAS is 46 m long, 25 m high, and 25 m wide. The detector characteristics were specifically engineered to enable a wide range of experimental work, although its initial task will be to search for the Higgs particle (see Chapter 3). At the heart of the detector is the nested, barrel calorimeter system (Fig. 2.19), and around this is a massive superconducting, toroidal magnet assembly (Fig. 2.20). Each end of the ATLAS experiment is capped with a fan-like muon detector system (Fig. 2.21).

The Compact Muon Solenoid (CMS) is located diametrically opposite the ATLAS experiment chamber (Fig. 2.22), and while it, too, will also be looking for the Higgs particle its experimental setup is entirely different from that adopted in ATLAS. The CMS detector is built around a massive, superconducting solenoid – a cylindrical coil that will generate a massive 4 T magnetic field. Coming in at about half the length of ATLAS, the CMS experiment is some 21 m long, 15 m wide, and 15 m high.

A transverse slice through the CMS is shown in Fig. 2.23, where the nesting of the inner calorimeters and the muon detectors are clearly seen. The muon detectors (Fig. 2.24) are composed of interleaved plates of iron (shown as dark grey blocks in Fig. 2.23) and detector chambers; it is the iron yoke component of the muon detector that contributes to the staggering 12,500 ton total mass of the CMS. One of the key design characteristics of the CMS is its ability to measure muon paths to a very high order of accuracy. This is required since the "events of interest" signaling the formation of a Higgs will have very specific muon ejection characteristics, and it is for this reason that the data-gathering trigger is hard-wired (in part) to what the muon detectors record (Fig. 2.25).

The ATLAS Experiment

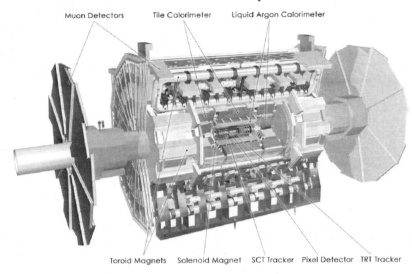

Muon Detectors   Tile Calorimeter   Liquid Argon Calorimeter

Toroid Magnets   Solenoid Magnet   SCT Tracker   Pixel Detector   TRT Tracker

Fig. 2.18 Schematic layout of the ATLAS detector. (Image courtesy of CERN)

Fig. 2.19 Engineers installing and aligning the inner tracking and calorimeter system located at the heart of ATLAS experiment. (Image courtesy of CERN)

Fig. 2.20 Looking down the bore sight of the ATLAS detector. The series of 8 "tubes" running lengthwise through the detector form part of the superconducting toroidal magnet system. The researcher at the lower center of the image provides some idea of the massive scale of the detector. (Image courtesy of CERN)

Since accurate tracking, to of order 10–100 μm, is one of the important characteristics of both the CMS and ATLAS experiments, the design engineers have included a whole host of system monitors. In ATLAS there are some 5,000 optical alignment sensors to monitor the movement of its various components. There are also 1,800 magnetic field sensors within the detector to enable researchers to map out and monitor the magnetic field strength and its configuration.

All of these system parameters will need to be known in order to accurately reconstruct particle tracks. The CMS magnetic field should be much more uniform than that produced inside of the ATLAS detector, and it is accordingly only monitored by some 80 sensors. The detector geometry itself, however, is measured by some 1,400 sensors using an internal light-based reference system.

Not only is the geometry of the detectors constantly measured, but so, too, are corrections relating to the phase of the Moon added in during the final analysis. Indeed, the Moon's motion raises tides in Earth's crust (just as it raises tides in its oceans), and the flexing of the bedrock within which the 26.6-km LHC ring system is housed will result in a 1-mm variation in its circumference according to whether the Moon is either in its full or new phase. In the everyday world such a change is unnoticeable. In the collider world, however, a 1-mm change in the confinement ring circumference will result in a measurable change in beam energy.

Last, but by no means least, the TOTal Elastic and diffractive cross-section Measurement (TOTEM) experiment is located close to the CMS detector, and it will

Fig. 2.21 Engineers working
on one of the "big wheel"
muon detectors that cap each
end of the ATLAS detector.
(Image courtesy of CERN)

determine to exquisite accuracy the proton–proton interaction cross-section. The experiment is composed of a series of detectors housed in eight specially designed vacuum chambers called Roman pots. This odd name is derived from the fact that the housings for the experiments are vase-shaped, and because it was an Italian research group that first developed the design for the instrument in the early 1970s – Fig. (2.26). The pots will be set up either side of the CMS detector, and they will intercept protons that have been elastically, rather than collisionally, scattered by an encounter with another proton. Elastic collisions are those in which the total kinetic energy of motion is the same before and after the proton–proton interaction, while the cross-section area of interaction, as its name suggests, is a measure of the effective area for collisions.

Once the cross-section area is known over a range of energies it can be related back to the structural properties of the proton itself. Although TOTEM is one of the smallest instruments to be installed at the LHC, all of the other experiments (ATLAS, CMS, LHCb and ALICE) will use TOTEM's results to calibrate their

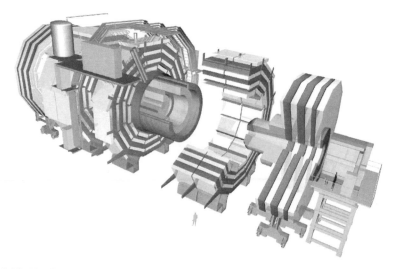

Fig. 2.22 The CMS detector. Note that the figures added to the schematic give some idea of the instrument's scale. (Image courtesy of CERN)

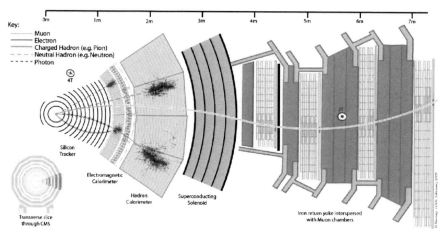

Fig. 2.23 Transverse cut through the CMS detector showing its multi-layered detector system. (Image courtesy of CERN)

luminosity monitors. The luminosity is directly related to the number of particles per unit area per second at a detector's collision point, and it is of great interest to the researchers since it relates to the probability of a rare collisional event, such as Higgs production, occurring. The greater the luminosity the more likely it is that a rare event will be detected, since the particle interaction rate is the product of the luminosity times the cross-section area of interaction.

The stage upon which new science might appear is now set, and the audience is seated and waiting. The LHC has been assembled, the systems have been checked and counter checked countless times, and the experiments are primed and ready to

Fig. 2.24 One of the CMS detector's muon detector end caps. The central core shows part of the hadronic calorimeter. The technician provides a clear indication of the detector's scale. (Image courtesy of CERN)

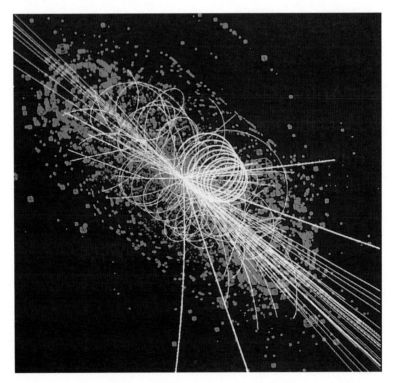

Fig. 2.25 A simulated Higgs event in the CMS detector. In this particular simulation the Higgs decayed into two Z particles, which further decayed into four muons. The four muon tracks "exit" the image half way along its bottom and right hand sides. (Image courtesy of CERN)

Fig. 2.26 Technicians work to install one of the "Roman pots" that will be used in the TOTEM experiment. (Image courtesy of CERN)

deliver a literal computer-disk mountain of data. With little doubt, we are certain to be amazed at what will be found. Much of the initial drive to build the LHC was centered on the desire to discovery the Higgs particle, an entity whose existence was first suggested by Peter Higgs (and independently by several other researchers) in the mid-1960s. In the next chapter we shall explore the reasons why the Higgs is such an important particle (indeed, with a great deal of hyperbole it has been called, by some, the "God particle") and what its discovery, or non-discovery, will mean for the Standard Model of particle physics.

Chapter 3
The Standard Model, the Higgs, and Beyond

Generation the First – An Acrostic

The making of matter,
Here-in beginning:
Emerging from darkness; strings and abstractions,
Munchausen dots in the void.
Arranged in fare groupings, and
Keenly selected
In time to appear amid clouds and deep clusters,
Nuclear bundles – three quarks in the making.
Great mountains are shaped and galaxies formed
Of up and down fractions invented by Gell-Mann
Forever entrapped; in hadrons be-jailed.
Make thee the neutron and proton in bulk
And add to the mix electrons all spinning
There then sits matter, the indomitable Thing,
Tangible and listless, worlds all creating,
Empty and open, yet bursting with promise,
Rendered in form by Pauli endorsing.

Take a look at Fig. 3.1. It is remarkable. Just as a poem can be thought of as the distillation of numerous thoughts and emotions all jostled into a minimalist collection of just the right words (sadly, as illustrated above, not a skill the author can claim), so Fig. 3.1 is a visual poem to the Standard Model of particle physics; it is the apex, indeed the shining pinnacle of a massive pyramid underlain by mathematics, experimental physics, extraordinary human intuition, and breathtaking genius. Within Fig. 3.1 is a description of all known matter – the very stuff of the tangible universe. As discussed in Chapter 1, all of the ordinary matter out of which we, this book, the chair you are sitting on, Earth, the stars, and the visible cosmos are made up of are themselves made up of just three first-generation fundamental particles: the up (u) and down (d) quarks and electrons. The electron neutrino rounds out Generation 1 as a Rip Van Winkle, near-massless, ghost-like particle. All first-generation particles, as far as we know, are stable (possibly forever?) and indivisible; they are truly the fundamental building blocks.

The second- and third-generation particles are also composed of quarks – the charm and strange, as well as the top and bottom, along with the muon and tau

M. Beech, *The Large Hadron Collider*, DOI 10.1007/978-1-4419-5668-2_3,
© Springer Science+Business Media, LLC 2010

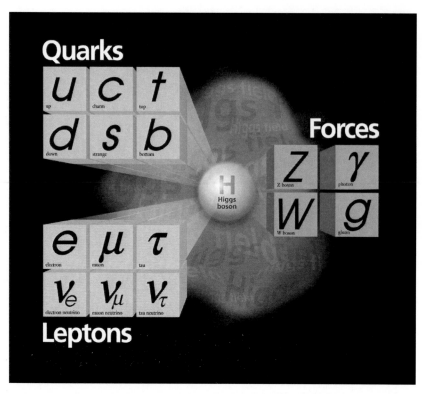

Fig. 3.1 The Standard Model of particle physics. All possible forms of matter are made up of quarks and leptons. The *right-most box* shows the force-carrying particles, while in the *middle* of the image lurks the yet-to-be-found Higgs boson. (Image courtesy of Fermilab)

leptons. Generations 2 and 3 are ephemeral and have no place in our everyday world. The particles made from these generations exist on only the smallest of timescales, a time much shorter than the blink of an eye, before they decay into a spray of stable Generation 1 particles. In essence Generations 2 and 3 are higher-mass copies of the Generation 1 particles, although it is presently not known why this structure repeats itself or why there are just three generations.

Figure 3.1 reveals that, just as there were 12 apostles that carried the good word into the world, so there are 12 particles that can create all the matter that we know of and that we can presently study in the laboratory. The picture is a little more complicated than this analogy, however, in the sense that each of the 12 particles has an associated anti-particle – identical in every way except for charge. (Since Christian doctrine does not include the idea of 12 anti-apostles, our analogy breaks down at this point.)

The Standard Model, although highly successful, is not a complete theory; we know that it is not a full description of matter in terms of Plato's ideal forms (recall Fig. 1.4). The most important open question at this time, and indeed, the primary scientific reason for building the LHC, is how do the various particles, from whichever generation, acquire their mass? Why does a tau, for example, have a mass 3,327 times greater than that of the electron and so on?

Likewise, the Standard Model predicts that all neutrinos should be massless, but as revealed in Chapter 1, this is now known to be untrue; neutrinos have mass[1]. The existence of dark matter (see Chapter 5) and the possible existence of dark energy (see Chapter 6) also indicate that the Standard Model requires expansion. As with all scientific models and statements, however, with the Standard Model we are really just talking about the best current model that answers the majority of questions that have been posed to date. We know, perhaps hope is a better word, that the Standard Model is probably not too far away from being right, but we also know that it must eventually be adapted and expanded to contain many more subtleties, particles, and complications.

The development of the Standard Model builds upon more than a 100 years of discovery, experimentation, and theoretical development. The story began with the discovery of the electron by British physicist J. J. Thomson in 1897, followed up shortly thereafter by the experiments conducted by Rutherford to reveal the structure of the atom in the early 1900s (see Chapter 1). The neutron was first identified at the Cavendish Laboratory, Cambridge, by James Chadwick in 1932, with the muon being discovered by Carl Anderson, who was studying cosmic ray events (see Chapter 7) in 1936.

The first experimental detection of the electron neutrino was announced by Frederick Reines and Clyde Cowan in 1956, with the discovery of the muon neutrino being presented by Leon Lederman and co-workers in 1962. The first experimental indication for the existence of up and down quarks was collected at the Stanford Linear Accelerator Center (SLAC) in 1969, with the charm and bottom quarks being identified at SLAC and Fermilab (Fig. 3.2) in, respectively, 1974 and 1977. The tau lepton was the next particle to be detected, confirmation coming from Martin Perl and collaborators at SLAC in 1976. Working with the Super Proton Synchrotron (SPS; see Fig. 2.8) at CERN, Carlo Rubbia, Simon Van der Meer, and hundreds of co-workers announced the discovered the W and Z bosons in 1983, while researchers working with CERN's Large Electron Positron (LEP) experiment were able to demonstrate, in 1991, that there can only be three neutrino types (that is, three generations of particles). The top quark was identified by researchers at Fermilab in 1995, and this success was followed shortly thereafter with the discovery of the tau neutrino in 2000.

The quark and lepton types identified in Fig. 3.1 are fundamental in the sense that they have, or at least are believed to have, no substructure. All other matter is made up of combinations of quarks and leptons. As described in Chapter 1, all of the stable matter in the universe is made of atoms that are composed of protons,

---

[1] There is a physics joke here (well, some call it humor) that runs along the lines, when informed that neutrinos have mass the confused Professor of Divinity commented that he hadn't even known they were Catholic – insert laughter or groan at this point.

Fig. 3.2 The Fermi National Accelerator Laboratory (Fermilab), Batavia, IL. Founded in 1967, this facility now houses the Tevatron particle accelerator with its 6.28-km confinement ring (seen in the background) in which collisions between protons and antiprotons are engineered. (Image courtesy of Fermilab)

neutrons, and electrons, with the protons and neutrons being made up of three up and down quarks apiece.

The various subgroups of matter are identified according to how many and which type of quarks they contain. All particles made up of quarks and antiquarks are known as hadrons, and the hadron group contains the baryons, which are particles composed of three quarks (such as the proton and neutron) and the mesons, which are composed of quark and antiquark pairs. Pions are members of the meson group, with, for example, the π^0 being composed of a combination of an up and antiup (written as u$\bar{\text{u}}$) quark-antiquark pair. The key additional point about hadrons is that they are all held together by the strong nuclear force, which is mitigated through the exchange of gluons.

Feeling the Force

The right-hand box in Fig. 3.1 indicates the various force-carrying particles (called bosons) that are incorporated into the Standard Model. The Z and W, so called intermediate vector bosons, mitigate the weak nuclear force that is responsible for radioactive decay. The gluon (g) is responsible for transmitting the strong nuclear force that binds quarks together, and the photon (γ) is associated with electromagnetic fields.

The strong and the weak nuclear forces only act over very small distances – smaller than that of the atomic nucleus, and are accordingly unfamiliar to us in the macroscopic world. The electromagnetic force, on the other hand, operates over distance scales much larger than those of the nuclear forces. Indeed, the photon, a localized vibration that propagates in the electromagnetic field, may in principle travel across the entire universe.

Although not a component of the Standard Model, a fourth fundamental force in nature, gravity, operates over extremely large distances. Indeed, it is gravity that shapes the very matter distribution of the greater universe. On the atomic scale, however, gravity is completely negligible. Indeed, if one compares the gravitational force between a proton and an electron located within the ground state of a hydrogen atom (recall Fig. 1.13) to that of the Coulomb force related to their opposite electrical charges, then the gravitational attraction is some 10^{-40} times smaller and therefore entirely negligible. On the large scale gravity is, as we shall see in Chapter 4, described according to the curvature of spacetime and Einstein's theory of general relativity. On the very smallest of scales, and at the very highest of energies, which must have prevailed in the very early universe, gravity must, as far as we can speculate, have behaved in a quantum-like fashion, and in this setting the gravitational force would be carried by gravitons. To date there is no universally accepted theory of quantum gravity, which is not to say that there are not a very large number of candidate theories all vying for recognition – as we shall see later in this chapter.

The strong nuclear force acts upon quarks and is responsible for holding particles such as protons and neutrons together. Unlike the familiar gravitational force of our everyday world, however, which is always attractive, the strong force is both repulsive and attractive, according to distance. If two quarks, for example, approach each other too closely, then the gluon exchange becomes repulsive, forcing them apart; if their separation becomes overly large, then the gluon exchange is attractive, bringing them back together. In this manner the quarks are always confined and continuously interacting with each other via gluon exchange (Fig. 3.3).

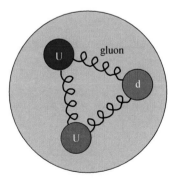

Fig. 3.3 The quark structure of a proton. A quark can posses one of three possible color charges (*blue, green*, or *red* – antiquarks carry anticolor charges). A quark carrying one color can be bound to another quark carrying its anticolor (making a meson), and three quarks, each with a different color charge, can be bound together (making a hadron)

The way in which all these complex interactions work is described by a theory known as quantum chromodynamics (QCD), which ascribes a color charge (or flavor) to quarks and gluons. Unlike electrical charges, which are either positive or negative, there are eight possible color flavors that can be ascribed to a gluon.

The picture of proton structure is actually a lot more complicated than suggested by Fig. 3.3. It turns out that virtual quark + antiquark pairs can be created within the proton by quantum mechanical vacuum fluctuations. This process is allowed in the quantum world provided that the energy borrowed to create the virtual quark pairs is repaid rapidly by their annihilation. This is just the Heisenberg uncertainty principle at work with the constraint that the small amount of energy borrowed ΔE is replaced in a time Δt such that $\Delta E \, \Delta t > \hbar$.

We have already seen that the proton is composed of three quarks, but if one adds up their total mass then the sum accounts for only about 1% of the proton mass – the remaining 99% of the proton mass is made up, it turns out, of the energy associated with vacuum fluctuations (remember Einstein's famous equation $E = m \, c^2$ tells us that energy and mass are equivalent and related by the speed of light squared).

To prove this result is an extremely difficult task, but it has recently been accomplished, for the first time, by a research group working at the John von Neumann Institut für Computing in Jülich, Germany. Writing in the November 21, 2008, issue of *Science* magazine Stephan Dürr and a team of 11 co-workers described how they tackled the problem of determining the mass of the proton and neutron ab initio by allowing for the formation of virtual quark + antiquark pairs in what is called a lattice QCD calculation. It required more than a year to complete the calculations using a parallel super-computer system performing 200 trillion arithmetic calculations per second. The results, however, determined the proton and neutron masses to within 2% of their measured values – a remarkable calculation. Incredibly, the calculations indicate that the basic building blocks of atomic nuclei attain the vast majority of their mass from the energy associated with the formation of virtual quark-antiquark pairs along with additional gluons bursting briefly into existence from the quantum vacuum. For all its mathematical sophistication, the computations by Dürr and co-workers are still not exact, and one of the key ingredients that the calculations didn't include is that of the Higgs field. As we shall see shortly, it is the Higgs field that gives particles their actual mass, and it, too, has an associated particle, called the Higgs boson, which will also be created, fleetingly, out of the quantum vacuum. We shall have more to say about the quantum vacuum in Chapter 7.

The weak nuclear force is responsible for describing the process of nuclear decay and transmutation – such as the beta decay in which a neutron is converted into a proton along with an electron and antielectron neutrino. The weak force operates through three so-called intermediate vector bosons: W^+, W^-, and Z^0, where the +, – and 0 correspond to their associated electrical charge. The weak force operates on scales much smaller than that of the strong force, but as its name suggests it carries relatively less strength than the strong force. The importance of the weak force is embodied in the fact that it can change the color flavor of a quark, and it is this property that results in particle transmutation. The beta decay, for example, is mitigated according to the conversion of a down quark into an up quark via the emission of a W⁻ vector boson.

Remarkably, our very existence is tied to the operation of the weak force. As we saw in Chapter 1 stars, our Sun being no exception, are stable against rapid collapse because they have hot interiors, and the energy to keep the stars hot and to stop them from cooling off and collapsing is generated through nuclear fusion. In the Sun this process works through the proton–proton (PP) chain, which results in the conversion of four hydrogen atoms into a helium atom with the release of energy (4H $\Rightarrow$ He + energy). The details of the PP chain are shown in Fig. 3.4.

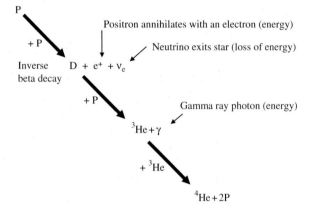

Fig. 3.4 The proton–proton chain by which four protons are converted into a helium atom with the release of energy. The first step in the chain requires the two protons to be close enough to interact and for one of the protons to undergo an inverse beta decay at the same time to produce a neutron

The first step in the PP chain is to produce deuterium, which has a proton and a neutron in its nucleus. This might at first seem odd since the process starts with two protons coming together, but nature doesn't allow the formation of helium-2, that is, a stable two proton nucleus. The only way in which the PP chain can move forward, therefore, is if one of the protons at exactly the right moment undergoes an inverse beta decay (a process governed by the weak nuclear force) to become a neutron. In this manner the neutron and second proton can then combine to make the deuterium nucleus D, and the W⁺ that mitigated the proton transmutation decays into a positron and electron neutrino – as shown in Fig. 3.4.

We shall have more to say about the nuclear timescale upon which stars evolve in Chapter 7, but for the moment we note that the timescale over which the PP chain can operate in the Sun is of order 10 billion years. This long timescale only comes about because of the very slow first step required to produce deuterium. If nature allowed helium-2 to be a stable nucleus and the PP chain didn't have to produce deuterium in order to proceed, then the Sun would only shine for a matter of years rather than billions of years. This is clearly crucial to our existence, since it has taken evolution 4.56 billion years to get life on Earth to where it is today, and this has only been possible because Earth has been warmed and nurtured by the near constant energy output of the Sun over the same timescale. Remarkably, we exist because of the weak interaction – within the macrocosm is the microcosm.

The Higgs Field – Achieving Mass

The Standard Model, while highly successful in describing the observed particle zoo, does not provide us with any direct insight as to why the various particles have their observed masses. Since this is such a fundamental point, physicists have, for many years now, been working towards developing a theory that literally endows the universe with mass. The current, best contender, for this theory invokes what is known as the Higgs field – named after Professor Peter Higgs (Emeritus, University of Edinburgh, Scotland; Fig. 3.5), who along with Robert Brout and François Englert (Free University of Brussels, Belgium) first developed the idea in the mid-1960s.

Fig. 3.5   Professor Peter Higgs. A portion of the theory relating to the Higgs field is shown on the chalkboard. (Image courtesy of CERN)

What has become known as the Higgs field was initially developed as a purely mathematical construct in an attempt to solve the mass generation problem. It is still unproven, but may soon be promoted to a quantified component of the Standard Model. Indeed, the primary motivation for constructing the LHC is to search for, and verify, that the Higgs field and the associated Higgs boson really do exist.

Although now in his early 80s, Higgs still recalls the whirlwind few weeks, starting Thursday, July 16, 1964, when the original idea first came to him, and during which the first two very short papers on the topic were rapidly written. Higgs recalls that he was sitting in the university library mulling over a research paper that had just been published by Walter Gilbert (Jefferson Laboratory of Physics, Harvard University). The paper presented a viewpoint with which Higgs strongly disagreed, and he knew that a rebuttal was required – but how? Gradually Higgs realized, as he thought more about Gilbert's vexing paper that an idea for an answer was evolving in his mind; he realized how elementary particles could acquire mass, and the rest, as they say, is history. The Higgs field was born.

Ironically, the first journal that he submitted his research paper to, the CERN-based journal *European Journal of Physics,* rejected it on the grounds that it didn't display any indication of being an important physical result. Dejected and not well pleased at having his paper returned, Higgs added just a few additional paragraphs to his article, and then submitted it to the journal *Physical Review Letters,* where it was accepted for publication on August 31, 1964. The paper was just two journal pages long, but it was destined to set the world of particle physics alight.

What is the Higgs field? Technically it is what physicists call a scalar field, which assigns a specific scalar value (that is, a real numerical quantity) to every point in space. In this manner the Higgs field permeates all space, and through its interaction with the fundamental particles it provides those particles with a mass. Any particle that doesn't interact with the Higgs field will be massless. So, for example, photons don't interact with the Higgs field (they are genuinely massless) but electrons do, since they have a measured (rest) mass of 9.10938×10^{-31} kg. Former director of Fermi Lab and Nobel Prize winner Leon Lederman (Illinois Institute of Technology) has likened the effect of the Higgs field to that of running swiftly over solid ground versus knee-deep in oil. If a particle, such as a photon, has no interaction with the Higgs field then it will move swiftly and remain massless; this is the running on solid ground part of the analogy. A particle that does interact with the Higgs field, however, will be slowed and act as if it has gained mass in the same way that a runner will be slowed down (as if their mass had increased) when trying to run through a deep oil slick.

Pushing the analogy to its limit, the bigger the runner's feet (i.e., the stronger the interaction with the Higgs field) the slower their pace and the greater their apparent mass. In Lederman's analogy, the Higgs field is like an oil slick that pervades the entire universe. Again, pushing the analog to its limits, just as one might throw a stone into an oil slick to set up a wave-like vibration, so, too, can the Higgs field be excited to vibrate. Such vibrations will be manifest as the Higgs boson, and the Higgs boson will "carry" the Higgs field in the same way that a photon "carries" the electromagnetic field. Finding evidence for the formation and then decay of the Higgs boson is the primary goal of the initial LHC research program – it is the *experimentum crusis*. Indeed, both ALICE and CMS will be looking for the Higgs through the independent study of its possible decay modes.

Feynman Diagrams

The Higgs boson cannot be observed directly; only the ghosts of its presence can be detected. In the standard picture the Higgs boson, once formed, will rapidly decay into other elementary particle products, but fortunately for the experimental physicists the decay modes should be very specific. Schematically the process runs as follows:

$$\text{P} + \text{P } \textit{collision} \Longrightarrow \textit{Higgs} \Longrightarrow \textit{Decay products} \Longrightarrow \textit{Observations}$$

In this manner it is envisioned that two protons collide (technically it is two of their quarks that collide) to produce a Higgs particle, which then decays into other more stable particles that can then be identified and tracked in a detector.

The problem for the experimental physicists, as we shall see below, is that the dominant decay modes will depend upon the precise value of the Higgs particle mass – and this is currently not known. Indeed, determining the mass of the Higgs is one of the key results expected to come out of the LHC experiment. It is the requirement to measure and track the particles produced in the various Higgs decay modes that has driven the designers of the LHC experiments (as described in the last chapter) to build highly sensitive calorimeters and precise tracking systems into their detectors. The rapid identification and tracking of particles will be important to the success of the program, and indeed, the key function of the massive computing system at CERN is to make sense of the 25 or so specially selected collision events recorded per second while the LHC is in operation.

The main tool for recreating particle decay chains and interactions was developed by Richard Feynman in the 1950s (see Fig. 3.6). Indeed, the so-called Feynman diagrams show how particles can interact and combine, decay, and be transmuted into new particles. By recording the detector tracks and identifying which particle types are present in a given interaction the vast computer mill at CERN can recreate the Feynman diagram and search for key signatures that indicate the production of a Higgs boson.

Feynman diagrams are really mathematical shorthand, and in many ways they are the visual poems that describe complex particle interactions. Figure 3.7, for example, shows the Feynman diagram of two electrons scattering through the exchange of a photon. The diagrams are constructed according to a very specific notation and essentially show the temporal process as the particles approach, interact, and then move apart again.

Fig. 3.6 Nobel Prize-winning physicist Richard Feynman (1918–1988) photographed during a presentation at CERN in the 1970s. (Image courtesy of CERN)

Fig. 3.7 Feynman diagram
of two electrons scattering via
the exchange of a photon. The
diagram can be read
according to the time
increasing towards the right

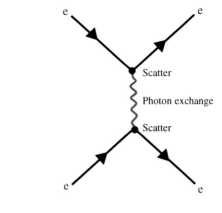

Fig. 3.8 Feynman diagram
illustrating annihilation and
pair creation. The convention
is to use *solid arrows* for the
trajectories of matter particles
and *open arrows* for
antimatter particles

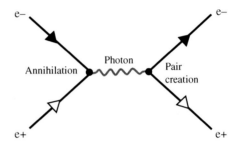

An example of a matter-antimatter annihilation event, followed by a pair creation event, is shown in Fig. 3.8. In this case an electron and positron interact and then annihilate to produce energy in the form of a photon. The pair creation event illustrates the reverse process where energy (in the form of a photon) is converted into an electron plus a positron pair.

The beta decay process in which a neutron decays into a proton, an electron, and an electron neutrino n $\Rightarrow$ p + e + v (the process studied by Fermi to predict the existence of the neutrino, as described in Chapter 1) is succinctly illustrated by a Feynman diagram (Fig. 3.9). Here, one of the down (d) quarks in the original neutron is converted into an up (u) quark through the emission of a W$^-$ boson, which is mitigated through the weak force, and this in turn decays into an electron and an antielectron neutrino.

The strong force interaction between two quarks can also be illustrated by a Feynman diagram, and Fig. 3.10 shows a scattering event mitigated through the exchange of a gluon. This diagram, as one would expect, is similar to that for the electron scattering event shown in Fig. 3.7.

The key, first analysis step for LHC researchers will be to use the vast computing power available to them to construct Feynman diagrams for the experimentally observed particle tracks, and then from these diagrams infer the possible generation of a Higgs boson somewhere in the decay chain.

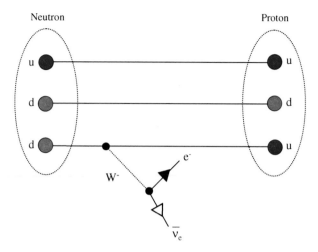

Fig. 3.9 Feynman diagram illustrating the beta decay of a neutron into a proton

Fig. 3.10 Feynman diagram
of a quark + quark scattering
event through the exchange of
a gluon

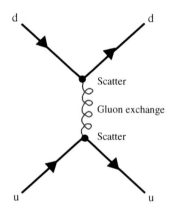

Searching for the Higgs

> For the Snark's a peculiar creature, and won't
> Be caught in a commonplace way.
> Do all that you know, and try all that you don't:
> Not a chance must be wasted to-day!
>
> The Hunting of the Snark – Lewis Carroll

Does the Higgs field really exist? At present we do not know for certain, but
once the LHC has been running for a few years we will know for sure – one way or
the other. Ever a person to think beyond the ordinary, however, Stephen Hawking
has recently placed a bet to the tune of $100 that the Higgs won't be found. We will
review the reasons for Hawking's, as well as other physicists, doubts shortly. For the
moment let us concentrate on what the standard picture has to say about the Higgs.

The main problem that the CERN researchers will initially face in the search relates to the fact that there are a large number of ways in which the Higgs might be produced and then decay. In addition, the dominant decay mode will vary according to the currently unknown mass of the Higgs itself. There is also the issue of background confusion of other, more mundane decay events, swamping the rare Higgs signal. The search that the LHC researchers face is much worse than that of the proverbial needle in a haystack.

In the low mass Higgs domain, where $M_{Higgs} < 130$ GeV, the most promising discovery channels relate to the decay modes producing either a pair of photons or a pair of τ leptons: $H \Rightarrow \gamma\gamma$ or $H \Rightarrow \tau\tau$. Under these circumstances the CMS detector (Fig. 2.22) will rely upon its sensitive electromagnetic calorimeter (ECAL) system to look for the $H \Rightarrow \gamma\gamma$ decay mode (Fig. 3.11), while the ATLAS (Fig. 2.18) detector will exploit its high performance hadronic calorimeter to investigate the $H \Rightarrow \tau\tau$ and subsequent decay modes. In the high mass Higgs domain it is possible that vector boson pairs can be produced ($H \Rightarrow WW$ and $H \Rightarrow ZZ$), and these will produce very specific decay signatures. The Feynman diagram for the $H \Rightarrow WW \Rightarrow$ $l\nu l\nu$ (two leptons and two neutrinos) channel is shown in Fig. 3.12.

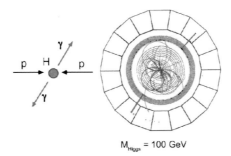

$M_{Higgs} = 100$ GeV

Fig. 3.11 Simulation of a Higgs event in the CMS detector. In this case the two photons generated in the $H \Rightarrow \gamma\gamma$ decay are recorded (the two *red bars* at the 1:30 and 7:00 o'clock positions) by the inner electromagnetic calorimeter (depicted end-on as the *green circle*). (Image courtesy of CERN and based upon the CMS brochure by CERN)

What has been dubbed the "golden channel" for the detection of the high mass Higgs ($M_{Higgs} > 130$ GeV) is the process by which four muons are produced. In this case, we have $H \Rightarrow ZZ \Rightarrow 4\mu$ and a simulated such decay within the CMS detector is shown in Fig. 3.13. In this case it will be the high performance and excellent tracking ability of the CMS muon detection chambers that will be the workhorse of discovery.

While the search for Higgs decay signals will be time consuming and complex, at least the researchers know in principle what it is they are looking for. By gathering in as much data as possible and seeing which decay channels are present and how often they occur the researchers will eventually be able to pin down the mass of the Higgs boson (provided, of course, that it really does exist).

Fig. 3.12 Feynman diagram
of the H ⇒ WW decay
channel in which the Higgs
decays into W⁺ and W⁻
vector bosons, which then
decay into two leptons (l) and
two neutrinos (ν). (Image
courtesy of Fermilab)

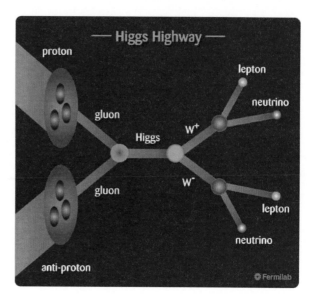

Fig. 3.13 Simulated Higgs
production event within the
CMS detector. In this case the
H ⇒ ZZ ⇒ 4 muons decay
mode is illustrated. The muon
tracks are shown as the *dark
lines* extending into the
detector's outer muon
chambers. (Image courtesy of
CERN and based upon the
CMS brochure by CERN)

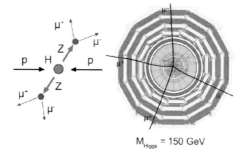

Peter Higgs has recently asserted that he is 90% confident that the LHC will find
his namesake particle, but others are less sure. Writing in the journal *Nature* for July
19, 2007, John Ellis (Physics Department, CERN) commented that "theorists are
amusing themselves discussing which would be worse; to discover a Higgs boson
with exactly the properties predicted by the Standard Model or to discover there is
no Higgs boson." In the first case, while the discovery would vindicate the Standard
Model, it would not provide researchers with any new insights into such topics as
dark matter (see Chapter 5) or supersymmetry (SUSY – to be discussed shortly). In
the latter case, the entire structure of the Standard Model would be cast into doubt,
and the great poetry illustrated in Fig. 3.1 would all be in vain. Not only would the
non-existence of the Higgs boson indicate that the Standard Model requires major
revision, it would also leave CERN in the somewhat embarrassing position of having

spent a huge amount of money on the LHC and having no new physics to show for it.

It is fair to say that this latter scenario is highly unlikely, but nature does have an alarming history of behaving in ways that are entirely unexpected. Stephen Hawking's $100 bet puts a positive spin to the scenario that would most alarm CERN diplomats by commenting that "It will be much more exciting if we don't find the Higgs. That will show something is wrong, and we need to think again."

Should the LHC researchers really be worried about the possibility of not finding the Higgs? The answer is probably no, but we shall just have to wait and see. In August of 2004, for a 2-week period only, the British firm Ladbrokes began accepting wagers concerning the likelihood of several scientific discoveries being realized before 2010. Among the topics on offer was the discovery of life on Titan, the detection of gravitational waves, and the discovery of the Higgs boson. The closing odds for the discovery of the Higgs before 2010 were an encouraging (for the LHC researchers) 6–1. Indeed these odds of discovery were much better than the 500–1 being offered for the detection of gravitational waves by 2010.

Well, even if the prize money wasn't going to give a good return on any investment, we now know that the Higgs won't be found before 2010, but, with the successful restart of LHC operations in November 2009 the verdict could well be in by 2012 or 2013. Those unlucky punters who missed the chance of a flutter on the Higgs during the Ladbrokes window of opportunity can still put their money down, however, and take a bet with Dr. Alexander Unzicker. Indeed, Unzicker, who teaches mathematics and physics in Munich, argues that not only will the Higgs not be found by LHC researchers but that it doesn't even exist. Placing a bet, argued German philosopher Emmanuel Kant, separates conviction from simple opinion, and those whishing to state their convictions can do so at http://www.bet-on-the-higgs.com/index.html.

There are some reassuring hints that the Higgs boson really does exist, and researchers working on the LHC's predecessor, the LEP collider, found candidate events that suggested the Higgs had been formed. In the LEP experiment a Higgs could be produced by a bremsstrahlung (breaking radiation) effect during the interaction of an electron and a positron. The data that the LEP researchers gathered was not definitive, but it did set a lower limit to the Higgs boson mass as 114 GeV – making it at least 100 times more massive than the proton.

More recent results coming out of Fermilab's Tevatron experiment are even more exciting. Although the Higgs has not been unambiguously detected with the Tevatron, the possible mass range for the Higgs has been constrained to be less that of about 160 GeV. Combining this upper mass cut-off with the earlier LEP results indicates that the Higgs, provided it does exist, must have a mass of between about 114 and 160 GeV (Fig. 3.14) – and this is right in the energy range that the LHC will be able to explore in great detail. If it's there, the LHC will find the Higgs.

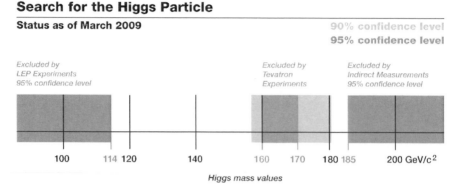

Fig. 3.14 Results released by Fermilab in March 2009 indicate that the Higgs boson cannot be more massive than about 160 GeV, while data from the LEP collider excludes masses smaller than about 114 GeV. (Image courtesy of Fermilab)

Supersymmetry

"Ignorance is strength," wrote George Orwell in his dystopian novel 1984, and while the context is entirely different from that of particle physics, the expression does encapsulate the notion that the less we know about a subject, so the greater is our freedom to speculate about it. Indeed, when it comes to thinking what might reside beyond the Standard Model theoretical physicists have not been shy in developing new ideas. Thousands, if not tens of thousands, of research papers have been published in the last 20 years on where the Standard Model might go next, but perhaps the most popular idea (not that popularity alone has anything to do with scientific correctness) is that of supersymmetry, or, as it is usually abbreviated, SUSY.

To see how SUSY works, we first need to look briefly at the quantum mechanical characteristics of the various particles in the Standard Model (Fig. 3.1). In Chapter 5 we saw how Niels Bohr developed Rutherford's atomic model (Fig. 1.13) and introduced the concept of quantized electron orbits. Specifically what Bohr quantized was the angular momentum of the electrons. What this means is that the angular momentum must be in integer units of Planck's constant.

Not only are the orbits of the electrons quantized but so, too, is the quantum mechanical property known as spin. Although the name suggests that spin is a measure of how fast a particle is spinning, it is actually a more complex concept and should be thought of as an intrinsic quantum mechanical characteristic of the particle. Given that spin is quantized, that is, expressed in units of Planck's constant h, it turns out that in nature there are two classes of particles: fermions and bosons. The key attribute of fermions is that they have half-integer spin values ($\pm$ h/2, $\pm$ 3 h/2, $\pm$ 5 h/2, ...), while bosons have whole integer spin values (0, $\pm$ h, $\pm$ 2 h, $\pm$ 3 h, ...). This spin characteristic difference may not seem like much, but it does, in fact, have a significant effect on how the particles can behave *en mass*.

Fermions (such as quarks, electrons and neutrinos), whose characteristics were first studied by Enrico Fermi and Paul Dirac (who we shall meet in Chapter 7) in the mid-1920s, are the loners of the particle world, and no two fermions can occupy the same position and quantum state at the same time. Indeed, we have encountered this rule earlier with respect to the electrons in the sense that they must behave according to the Pauli Exclusion Principle (Fig. 1.15). It is, recall, through the non-overcrowding condition of the exclusion principle that enables atoms to exist and chemistry to take place.

Bosons (such as all the force-carrying particles in the Standard Model, see Fig. 3.1) are the exact opposite of fermions and are thoroughly gregarious in nature, clustering together, the more the merrier, irrespective of location and quantum state. In the extreme of this clustering condition Bose–Einstein condensates can form, and these have many remarkable properties. The statistical behavior of integer spin particles was first studied by Indian scientist Satyendra Nath Bose (Dhaka University) in 1924, and this is how they acquired their name.

Where all the above discussion on spin and the behavior of fermions and bosons comes into play is that the equations that describe fundamental particle interactions should hold true if one switches a boson with a fermion. But the problem is they don't. The question for physics, therefore, is why, and the answer, at least the most studied possibility for an answer, is supersymmetry.

Under SUSY every particle in the Standard Model is paired with a more massive super-partner, called a sparticle; the fermions are pared with boson sparticles, and the bosons are paired with fermion sparticles. For experimental physics the SUSY prediction is absolutely delicious, since it says that only half of the elementary particle inventory has been detected to date. This being said, there is currently no experimental data that even hints at the possible existence of sparticles.

Under the SUSY paradigm the sparticle companions to fermions are given an added initial "s" in their name, so that the super-partner to the electron is the selectron, while those of the quarks are called squarks, and so on. The boson superparticle partners are given an "ino" ending, so that the gluon is paired with the gluino, and the Z boson is pared with the zino, and so on. Even the much sought after Higgs will have its supersymmetric counterpart, the Higgsino.

The basic mathematical theory behind supersymmetry was developed in the early 1970s, but the so-called minimal supersymmetry standard model (MSSM) was first described by Howard Georgi (Harvard University) and Savas Dimopoulos (Stanford University) in 1981. The MSSM is the most straightforward extension of the Standard Model, and it makes specific predictions about how sparticles might be experimentally observed. It is presently thought that the lightest supersymmetric particle (LSP) will be one of the four possible neutralinos, and some versions of SUSY (and there are many) predict that the mass of these sparticles will be in the TeV range, making them potentially observable within LHC detectors. Figure 3.15 shows a simulation for the outcome of a neutralino decay within the CMS detector. In this specific simulation the neutralino decays into a Z boson and an LSP, with the Z boson further decaying into two muons (shown as the solid lines to the left and lower right in the figure).

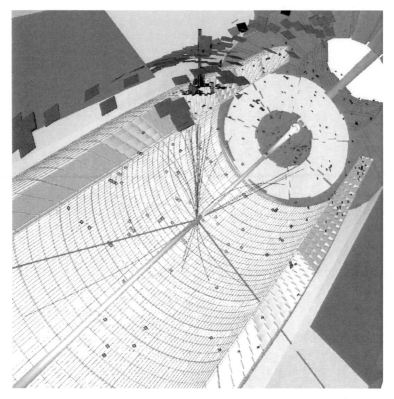

Fig. 3.15 Simulation of a neutralino decay in the CMS detector. The key features produced are the two muons, shown as the *solid lines* exiting to the left and the lower right and the localized energy deposition in the ECAL detector, shown as *dark squares* towards the center top of the image. (Figure courtesy of CERN)

Many supersymmetric decay modes are possible, but given time LHC researchers should be able to determine if any such decay signatures are present in their data set and thereby confirm, or possibly deny, the existence of SUSY.

Exotica: Going Up, Going Down

Theoretical physicists truly like the idea of supersymmetry. It is mathematically elegant, and it addresses some of the technical problems that are known to arise in the Standard Model. It also holds out promise for the construction of a viable grand unified theory (GUT) in which the fundamental forces of the subatomic world combine at very high energies. Indeed, one of the greatest triumphs of twentieth-century physics was the demonstration that at energies above about 100 GeV the electromagnetic and weak nuclear forces, which behave very differently at lower energies, become unified into a single, so-called, electroweak interaction.

Derived independently by Stephen Weinberg and Sheldon Glashow at Harvard University, and Abdus Salam in Trieste, Italy, the electroweak theory predicted the existence of three new particles, the W^+, W^-, and Z^0 -, and these particles, as we saw earlier, were successfully found by Carlo Rubbia, Simon Van der Meer, and co-workers at CERN in the 1980s.

The electroweak theory predictions and their experimental verification was a magnificent triumph for physics and human ingenuity, but the story hardly stops there. If the electromagnetic and weak nuclear theory unify at high energies, perhaps it makes sense to speculate that at even higher energies the electroweak and strong forces unify, making one super-interaction. Such grand unified theories make for highly interesting consequences in the field of cosmology, since the only time (ever) that energies have been high enough for GUTs to possibly operate was in the first instances of the Big Bang.

We shall describe the Big Bang theory in the next chapter, and simply note here that the present predictions are that the strong and electroweak interactions may have been unified for just the first 10^{-35}-th of a second of the universe's 14 billion year history. The electroweak interaction split into the weak nuclear force and the electromagnetic force once the universe had cooled to about a million billion Kelvin, at which time it was a stately one thousand billionth of second old. Such times and temperatures are unimaginably short and large respectively, and have absolutely no crossover with our everyday lives, but remarkably, it was within these minuscule moments and blistering temperature domains that the very physics of our universe was forged. Within the macrocosm is the microcosm.

Although the physicists and cosmologists evoke variously formulated GUTs as a means of explaining why the universe has its observed structure, there is still, incredibly, an even earlier epoch when presumably gravity combined with the strong and the electroweak interactions, resulting in one, original, super-interaction that saw the universe through the very zeroth instant of its birth. There is no universally accepted Theory Of Everything (TOE) to explain what happened prior to the Planck time of 10^{-43} s, but it appears – well at least some researchers argue so – that supersymmetry might pave the way to formulating a successful theory of quantum gravity when combined with the ideas of string theory.

With string theory our story becomes ever deeper, even smaller in scale and even more complex, but the basic idea is to treat quarks and leptons not as point, fundamental objects, but as two-dimensional strings, or in some versions of the theory multidimensional membranes. These strings and membranes, however, are minuscule, having dimensions of order the Planck scale: 10^{-35} m.

Since the early 1970s, when string theory in its present form burst onto the scene, it has witnessed an incredible growth in popularity, and it has taken many twists and turns in its formulation. Indeed, numerous versions of string theory have been presented, and while they all hint at ways of unifying the four fundamental forces they also invoke the existence of multidimensional space. It is not clear yet if 10, 11, or 26 dimensional space is required to make string theory work. Time, as ever, will tell if string theory is on the right track and whether it truly does explain, well, everything. We shall have a little more to say about the idea of extra spatial dimensions in

Chapter 7, since one possible outcome of this concept has resulted in a lawsuit being filed against CERN in an attempt to force the closure of the LHC. Incredibly, as we saw in Chapter 2, we find that even in the legal macrocosm is the atomic microcosm.

Finally, but by no means least, supersymmetry might provide an answer to one of the longest running problems of modern cosmology. Indeed, as we shall see in Chapter 5, it was realized nearly 80 years ago that there is much more matter in the universe than can be accounted for in the form of stars and gas (so-called baryonic matter). This dark matter component truly exists, as we shall see, but it is a form of matter that has mass and yet does not interact with electromagnetic radiation. Hence, it cannot be observed directly with any type of conventional astronomical telescope. Given the large implied masses, many astronomers and physicists believe that dark matter might actually be composed of sparticles. If the researchers at the LHC can confirm the existence of, say, the LSP, then not only will it have profoundly changed particle physics, it will have also profoundly changed our understanding of the universe. The future truly holds exciting promise, and the words of T. S. Eliot seem entirely apt to end this chapter.

> What we call the beginning is often the end
> And to make an end is to make a beginning.
> The end is where we start from.

Chapter 4
The Big Bang and the First 380,000 Years

Of old was an age, when was emptiness,
There was sand nor sea, nor surging waves;
Unwrought was Earth, unroofed was Heaven –
An abyss yawning and no blade of grass.

Völsungakvida en Nýja – J. R. R. Tolkien

Bearing witness to our running mantra that within the macrocosm is a reflection of the microcosm, the LHC will enable physicists not only to study the substructure of atoms. It will also enable them to investigate phenomena that haven't occurred in our universe for 14 billion years. Indeed, some of the events that will be studied at CERN haven't occurred since the very first moments that the universe sparked into existence. This is an incredible state of affairs, and while the LHC is not actually recreating the beginnings of the universe, or for that matter creating any new universes, it is allowing researchers to study the physical phenomenon that shaped and molded the very first fledgling instants of our cosmos.

Human beings, for so history tells us, have continually struggled to understand the greater cosmos. Indeed, the task of understanding the universe is far from an easy one, and many of the questions that can be raised are deep, profound, and extremely complex. Even if we sidestep the issue of origins, we are still left with the highly non-trivial problem of explaining why there is a universe containing any matter at all, and then, given that there is a universe, why it has the form that it does.

Prior to the early twentieth century there were no reliable astronomical data to indicate just how big the universe was. Indeed, before this time, with the absence of any firm observations about its contents and extent, the universe could be as large or as small as suited one's philosophy or religious bent. Although Nicolaus Copernicus (Fig. 4.1) speculated (correctly as it turned out) in his 1543 masterpiece *De Revolutionibus* that Earth orbited, as third planet out, a central Sun, he firmly set the boundary and extent of the universe as a spherical shell just beyond the orbit of Saturn.

The Sun was, according to Copernicus, located at the very center of the observable universe, and if Earth's distance from the Sun is taken as the unit of measure (the astronomical unit), then the universe was not much greater than 20 astronomical units across. For Copernicus the entire universe was a rather small and compact

M. Beech, *The Large Hadron Collider*, DOI 10.1007/978-1-4419-5668-2_4,
© Springer Science+Business Media, LLC 2010

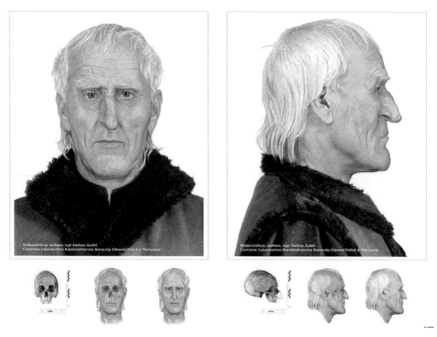

Fig. 4.1 Nicolaus Copernicus (1473–1543), Canon of Frombork Cathedral, Poland. The images shown are forensic reconstructions of Copernicus's face based upon the recent recovery of his skull in 2005

affair. Thomas Digges, on the other hand, soon pushed the Copernican model to an opposite extreme, and in 1576 placed the Sun and its attendant planets within the airy context of an unbounded stellar system (Fig. 4.2). The first stars in Digges's cosmology still appeared, however, just beyond the orbit of Saturn. Here is the classic dilemma of the cosmologist; two extremes of the universe can be constructed, one small, the other infinitely large, and both apparently accounting for all of the observations. The fundamental question, of course, is which of the two alternatives is the right one – or are they both wrong?

The history of science, of course, tells us that Copernicus and Digges were both right in the main part, but wrong concerning specific details. Copernicus was wrong about some of his assumptions concerning the motions of the planets, and Digges was wrong in suggesting that the universe is infinite in extent – which is not to say that it isn't very much larger than it might at first appear.

We shall trace the development of astronomical distance measures in Chapter 6, but for the moment let us simply note that reliable distances to the nearest stars were first obtained in the late 1830s, nearly 300 years after Copernicus first set them at rest in a sphere beyond planet Saturn. Within a 100 years of the first stellar distances being measured, however, the distances to other galaxies, remote "island universes" beyond the Milky Way, had also been measured, and a young Edwin Hubble was

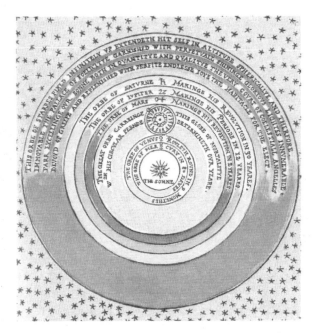

Fig. 4.2 The universe unbounded – the infinite stellar realm described by mathematician Thomas Digges in 1576

in the process of establishing the remarkable observational fact that the universe is expanding.

Hubble's observations will be examined in more detail in Chapter 6, where we discuss the more recent and rather mysterious discovery of an apparently accelerating universe. Our task in this chapter is to review the consequences of the universal expansion displayed in the motion of distant galaxies.

To briefly preempt our further discussion of Hubble's law, the key result that emerged in the late 1930s is that the further a galaxy is away from us, the faster is its velocity of recession; indeed, the speed and distance are linearly related by a constant H_0 – Hubble's constant. Importantly, the expansion is also found to be uniform – which, as we shall see, is crucial to our understanding of what is going on. By being uniform it is understood that all space is expanding at the same rate and that independent of where one might be located in the universe the expansion rate will always appear to be the same. This latter statement is often presented under the guise of the cosmological principle, which states that the universe is homogeneous and isotropic – that is, the universe looks the same everywhere and in every direction.

The cosmological principle carries a lot of unarticulated philosophical baggage, in part because it is a principle or working hypothesis rather than an absolutely proven statement of fact. (We shall come back to this issue in Chapter 6.) Importantly the Copernican principle also tells us that while it might look as though we are at the center of the universal expansion, as described by Hubble's law, we are, in fact, not centrally located. In essence, because the expansion is uniform it is as if every point in the universe resides at the center of the expansion. No matter where you are or where you go in the universe it will look as if all the other galaxies are moving away from you in accordance with Hubble's law. The cosmological principle also asserts that there are no physical boundaries or edges to the universe and that the universe isn't expanding, or gushing firework-like, from a central point into another space or region, but that it is simply getting bigger everywhere.

A commonly adopted, minds-eye picture used to describe the expanding universe is that of inflating a balloon with spots painted upon it. Think of the spots on the balloon as representing galaxies; then, as the balloon is inflated, its surface stretches and expands, moving the spots away from each other. Irrespective of which spot we choose as being the host galaxy of the observer, the expansion rate will be the same and uniform (assuming, that is, that we have a perfectly spherical balloon, and we ignore the air inlet tube; the problem with any analogy is that there are always caveats). In this analogy, since every point of the balloon's elastic surface is imagined to be expanding at the same rate, a Hubble-like law of expansion will come about.

Now, we must not think of the universe as being the entire 3-dimensional balloon in this analogy, but rather it is just the elastic surface of the balloon that represents our universe. The galaxies must not be thought of as all moving away from the center of our spherical balloon. Likewise, there is no initial center of expansion located on the surface. Remember, each and every point on our balloon's surface appears as if it is the central point of the universal expansion.

This also leads us to another important point evident in our analogy. Given that we are an observer located on the balloon's surface and that we observe a Hubble law-like recession of surrounding galaxies, this suggests that if we trace the motion backwards then surely everything will eventually collapse down to a single point. In other words there must be a unique center of the universe around which everything else is expanding – the primordial origin, formed, as it were, at time zero.

Again this is where our analogy has to be thought through carefully. With the balloon analogy of expansion the moment of universal creation, at time zero, does not correspond to a point but to the appearance of the deflated balloon's surface itself. It is only once the balloon's surface has appeared (in our case *ex nihilium*) that universal expansion sets in.

Perhaps the overriding caveat that goes along with our (indeed, with any) analogy, and with our observations of the real universe, for that matter, is that things are not always as they may at first appear. This certainly applies to the very early universe, those first stuttering moments when the potential for all that might possibly exist and come about over countless future eons came into being. Indeed, and in spite of all our current observations, physical theories, computer models, and mathematical skill we must not forget that the primordial breath is inherently unknown to us. We can speculate about what occurred in those very earliest of moments, indeed lack of ideas is not the problem; rather, the problem is one of constraint. Exactly how do we separate one very different theory from another, and exactly how do we make sense of the relatively few observational measurements presently available?

These issues, of course, are exactly what make cosmology one of the most dynamic and exciting branches of astronomy and physics.

However, things are perhaps not quite as bad as the paragraph above makes out. There are indeed several very clear observations that tell us about the state of the universe in its earliest moments of creation, and, of course, this is also where the LHC experiment will potentially provide new understandings relating to the forces, interactions, and elementary particles that existed when the universe was literally less than the blink-of-an-eye old.

The Big Bang

It is probably not unfair to say that modern cosmology began in 1916 when Albert Einstein introduced his first paper on general relativity; for indeed general relativity is our best current theory of gravity. Specifically, general relativity explains how gravity is a manifestation of the curvature of both space and time. And this, of course, is of interest to us in this chapter, since the dynamics of the universe are determined by gravity. Indeed, it is remarkable to think that it is gravity, the very weakest of the known fundamental forces (recall Chapter 3), that shapes the very largest of structures – the stars, the galaxies, clusters of galaxies, and the very cosmos itself. We will have more to say about the large-scale distribution of matter, galaxies, and galaxy clusters in the next chapter, but for the moment our attention will be focused on interpreting the essential meaning behind the general solutions to Einstein's equations.

One of the problems faced by a would-be theoretical cosmologist is that Einstein's equations of general relativity technically admit an infinite number of possible solutions. The equations as such have no unique solution, and they make no single prediction about the structure of the universe. At first take this situation might suggest that they are therefore next to useless in helping us to understand the universe. Perhaps this is partly true, but it does transpire that there are only a few of the possible infinite number of solutions that are actually interesting. Solutions that are ultimately uninteresting (apart from the pedagogical insight they offer), for example, are those in which the universe contains no matter, but only radiation; or universes that collapse after just half an hour of expansion, or universes that expand so rapidly that stars never form. Clearly, the solutions that are of most interest to cosmologists are going to be those that allow for the existence of matter and in which structures such as galaxies as well as clusters of galaxies can form and evolve, and in which a Hubble-like uniform expansion is possible.

The latter point is especially important, since if we do not allow for the continuous and spontaneous generation of matter, then we are also forced to accept that the universe we see around us had a starting epoch – a time zero as it were – and that it must, therefore, be of a finite age. The ever inventive, if not combative but always thought-provoking, British cosmologist Sir Fred Hoyle felt that such conclusions about the finite age of the universe were premature, and described the finite-age models as being Big Bang cosmologies.

Well, of course, when it comes to cosmology we should never say that we actually understand all that is going on. But it is fair to say that Hoyle's ideas (especially those relating to the steady-state model) are no longer considered viable by the vast majority of present-day researchers. In spite of his objections, Hoyle's Big Bang term has stuck as the general label for those cosmological models predicting a finite age and an early compact and very hot phase followed by expansion. Perhaps the Belgian mathematician Georges Lemaître best described these so-called primeval atom models when he wrote, "The evolution of the world [that is, cosmos] may be compared to a display of fireworks that has just ended; some few red wisps, ashes and smoke. Standing on a cooled cinder, we see the slow fading of the suns, and we try to recall the vanished brilliance of the origin of worlds."

All Big Bang cosmological models are based upon the premise that the universe began some finite time ago in a highly dense, high temperature state. Having accordingly come into existence the universe could then do essentially one of two things: it could continue to expand forever, or it could expand for a finite amount of time and then collapse back into its primordial atom-like state again. If the latter situation occurs, with the universe going through a collapsed end phase (appropriately called a Big Crunch), there are then no specific reasons why the universe might not "bounce" and start expanding all over again (Fig. 4.3). If a bounce does occur after a Big Crunch then, it has also been suggested, that the laws of physics and the values of the various fundamental constants (see Appendix 1) might also change.

The first mathematical solutions to Einstein's equations of general relativity, when applied on a cosmological scale, were published in 1922 by Russian physicist Alexander Alexandrovich Friedman. Working from within the besieged city of

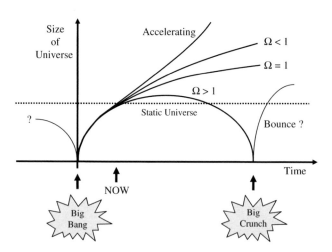

Fig. 4.3 After the Big Bang, the universe can either expand forever, or undergo a collapse and Big Crunch, and possibly even undergo repeated collapse and expansion cycles. The future of the universe is determined according to the parameter Ω, the ratio of the density of matter in the universe to the critical density

Petrograd (later Leningrad and now, once again, Saint Petersburg), Friedman developed his ideas during the 1917 Russian Revolution. Specifically, he studied the special family of solutions that require the cosmological principle to always hold true. These solutions are parameterized by two constants, H_0, Hubble's constant, and Ω, the ratio of the actual density of matter in the universe to the critical density (to be defined in a moment).

All of the possible Friedman model universes start off with a Big Bang, from a primordial atom state, full of energy, and they then undergo uniform expansion. The value of the parameter Ω, however, is crucial in that it not only determines the large-scale geometry of the universe (Fig. 4.4), it also determines its ultimate fate. If, for example, $\Omega < 1$ then the universe is said to be negatively curved and open. Under this geometry the universe will expand forever. If, on the other hand, $\Omega > 1$, then the universe is said to be positively curved and closed. Under this geometry all possible universes are transitory and will collapse to undergo a Big Crunch end-phase within a finite amount of time.

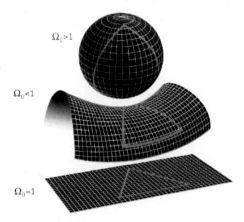

Fig. 4.4 Three possible geometries of the universe: open ($\Omega < 1$), closed ($\Omega > 1$), and flat ($\Omega = 1$). The inherent geometry of each possible universe is described according to the number of degrees contained within a *triangle* linking three points together. In our everyday world we are used to the fact that there are 180° in a triangle. This corresponds to the flat, so-called Euclidean, $\Omega = 1$, cosmology. When $\Omega > 1$, however, and the universe is closed, there are actually more than 180° in a *triangle*. In contrast, when $\Omega < 1$, and the universe is open, there are less than 180° in a *triangle*. (Image courtesy of NASA)

The special case, favored by many theoreticians for what are essentially aesthetic reasons, is when $\Omega = 1$. In this situation the universe is said to be flat, or Euclidian, and while the rate of expansion gradually slows down, it never actually stops (to thereafter undergo collapse) within a finite amount of time. A spatially flat universe is said to be Euclidian in honor of the ancient Greek mathematician Euclid of Alexandria. Indeed, in one of the most famous and influential books ever written, *The Elements*, compiled circa 300 B.C., Euclid describes the essential laws of mathematics and geometry.

Remarkably after twenty-one centuries Euclid's text still has relevance, and his mathematical proofs are still correct and true – as they will be, of course, for all eternity. This is a very different situation to that displayed by the history of cosmology, where new theories and paradigms appear, shift, and fall from grace on a near continual basis. The ever ambitious Galileo may well have argued during the early seventeenth century that the language of nature is mathematics, but the permanency of mathematical proof is something that is virtually impossible to achieve in the physical sciences. The universe is indeed more complicated than even the subtlety of mathematics can hope to explain.

The Critical Density and Ω

The critical density and the value of Ω are determined according to the amount of matter there is in the universe and how it is distributed. Specifically, the critical density is determined according to the so-called escape velocity, which in turn is determined by the amount of matter there is within a given region of space. The critical density is defined according to the Hubble expansion velocity being equal to the escape velocity, and Ω, as described earlier, is defined as the ratio of the observed density of matter in the universe to the critical value. Present-day observations indicate that the critical density comes in at about the equivalent of 5–6 protons/m^3. On the other hand the best surveys of the large-scale distribution of galaxies (the topic of Chapter 5) determine the observed density of matter to be equivalent of about 0.2 protons/m^3. These combined results indicate that $\Omega \approx 0.04 < 1$ and that the universe is open and will expand forever.

At first take it might seem that all is now solved and that the observations indicate that we live in an open Friedman universe that will expand forever. This, however, is not the case. As ever in cosmology, all is not as it might at first appear. Although we need not discuss further the numerical value of Hubble's constant in this section, we must make a clear statement about Ω, and specifically what the observed density of matter actually corresponds to. The statement made earlier that on the largest scales the density of matter is observed to be equivalent to about 0.2 protons/m^3 is based upon telescopic surveys of galaxies (including estimates for the interstellar and intergalactic gas and dust). In other words it corresponds to baryonic matter – the matter that can either emit or absorb electromagnetic radiation. Strictly speaking, therefore, we have really been considering Ω_B, where the subscript B stands for baryons.

If the universe just contained baryonic matter, then we would indeed have determined its properties as being open and set to expand forever. The reality of the situation, however, is that there is much more mass to the universe than can be detected with standard telescopes working with electromagnetic radiation. There are, it turns out, two major additional terms that contribute to the true value of Ω. Indeed, baryonic matter (stars, planets, gas, and dust) constitute only a minor fraction of the mass content of the universe, and it is dark matter (the topic of

Chapter 5) and dark energy (the topic of Chapter 6) that really determine the true value of Ω. As we shall discuss below and in Chapter 5 and 6 there are various reasons why cosmologists believe (one might better say hope) that we live in a universe in which $\Omega = 1$. One of the primary reasons for this belief is based upon another fundamental observation that clearly indicates that our universe must have begun with a Big Bang.

The Microwave Background

The discovery of the cosmic microwave background (CMB) is a classic example of having the right people in the right place at the right time with the right experimental equipment. The people involved were Arno Penzias and Robert Wilson, the place was Crawford Hill, New Jersey, the year was 1963, and the equipment was part of an experiment being conducted by Bell Telephone Labs to investigate satellite communications interference. In any radio-receiving system there is always inherent noise; some is due to the actual electronics, some is due to external sources, and some is due to the receiving antenna itself.

Since they were working with a state-of-the-art receiver, Penzias and Wilson were determined to make their system as noise-free as possible. Indeed, Penzias and Wilson were sticklers. The antenna horn (Fig. 4.5) was cleaned and scrubbed; even nesting pigeons were evicted from its interior. The antenna feed was pristine, and yet there remained an irritating hiss that just wouldn't go away. The noise signal was always there, day and night, rain or shine, and in every direction the antenna

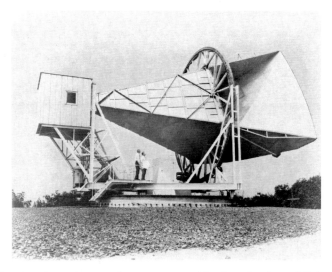

Fig. 4.5 The ultra-clean microwave horn antenna used by Arno Penzias and Robert Wilson to discover the cosmic microwave background

feed-horn was pointed. Further, the signal emission characteristics corresponded to those that would be expected from a blackbody radiator having a temperature of 3 K (a chilly −270°C).

Penzias and Wilson realized that the uniformity of the radio hiss across the sky couldn't be explained by discrete sources, such as local electrical interference or the Sun or the galaxy itself, and accordingly they reasoned that it had to be some kind of universal radio noise. But exactly what kind of source could it be? To find an answer they began by asking researchers at nearby Princeton University what astronomical sources might possibly have the characteristics of a 3 K blackbody radiator. The question stunned the Princeton cosmologists, and it was quickly realized that what Penzias and Wilson had stumbled upon was the relic radiation of the Big Bang itself. That such a relic should exist had, in fact, been predicted as early as 1948 by George Gamow, and his then student Ralph Alpha. Incredibly, the radio noise that Penzias and Wilson had tried so hard to eliminate from their system turned out to be the distant echo of photons released into the universe when atomic matter itself first began to form – at the time of recombination.

Once it was clear what the background radio hiss was due to, Penzias, Wilson, and now their Princeton co-workers began to map out the noise spectrum (Fig. 4.6). When the intensity of the radiation was plotted against wavelength, a perfect black-body spectrum emerged (recall Chapter 1). The detection of such a perfect spectrum clearly implied that the radiation must have been produced under conditions of ther-mal equilibrium – and the only time in which the universe would have been in such a state was far back in time, shortly after the moment of creation itself. It was an incredible result.

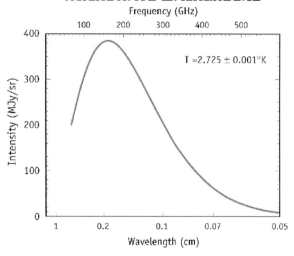

Fig. 4.6 The blackbody spectrum of the cosmic microwave background. The characteristic shape of the spectrum indicates that the temperature of the radiation is 2.725 K, and since the radiation intensity peaks at about 2 mm wavelength it falls in the microwave region of the electromagnetic spectrum

It was noted earlier that all Friedman cosmological models start from a high density, high temperature state (the Big Bang). Under these conditions all Friedman universes start off as boiling plasmas composed predominantly of photons, electrons, and protons with a smattering of helium nuclei. No neutral atoms can exist in the early phases of such a universe because the photons have so much energy that they will rapidly strip the electrons from any bound orbitals (recall Fig. 1.13).

Free electrons, however, are very efficient at scattering photons by a mechanism known as Thomson scattering (this is the same J. J. Thomson who developed the plum-pudding atomic model described in Chapter 1). The early universe would have the appearance, therefore, of a bright glowing fog of light. As the universe expanded, however, the fireball would begin to cool, and eventually the temperature would drop to a level below which stable atoms could start to form. At first the protons would capture free electrons to make hydrogen atoms, while other electrons would be captured by helium nuclei. (We shall discuss this process more fully shortly.)

At this time of recombination the freely moving electrons are removed from the Big Bang fireball and photon scattering becomes much less efficient – essentially the universe as far as the photons are concerned suddenly becomes transparent. What we now measure as the CMB corresponds to the radiation that escaped (that is, underwent a final scattering) at the epoch of recombination. Model calculations indicate that the time of recombination occurred some 380,000 years after the Big Bang. Originally, the radiation at recombination would have been in the optical and ultraviolet part of the electromagnetic spectrum, but the continued expansion of the universe has resulted in the downgrading (or more correctly stretching) of the radiation into the infrared and microwave parts of the electromagnetic spectrum – where it was found by Penzias and Wilson in 1964. For their serendipitous discovery Penzias and Wilson were awarded the Nobel Prize for Physics in 1978.

With the discovery of the CMB most cosmologists became convinced of the correctness of the Big Bang model. There was, however, a problem. The Planck spectrum (Fig. 4.6) of the CMB was too close to being perfect. The primordial fireball out of which the photons escaped at recombination was just too smooth and uniform, leaving no natural explanation for the origin and formation of large-scale cosmic structure.

The first detailed all-sky study of the CMB was completed with NASA's COsmic Microwave Background Explorer (COBE) satellite launched in 1989. Theoreticians had long believed that temperature fluctuations should be present in the CMB, and the COBE detectors finally revealed these small wrinkles (Fig. 4.7) at a level of about 10^{-5} K – small temperature variations indeed, but just enough to indicate that density inhomogeneities existed at the time of recombination, and enough variation to allow gravity to sink its teeth into and begin the process of building clusters of galaxies.

The Boomerang (Balloon Observations Of Millimetric Extragalactic Radiation and Geophysics) experiment, first launched in the late 1990s, offers a fine example of how scientists often go to extraordinary lengths to tease out nature's deepest secrets. In this case the highly sensitive bollometric detectors at the heart of the experiment were carried aloft on a high-flying balloon released into the Antarctic

Fig. 4.7 Small temperature variations appear as blobs and ripples in the all-sky CMB map produced from COBE satellite data. (Image courtesy of NASA)

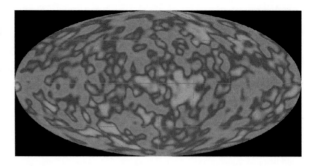

polar vortex. Accordingly, during each mission the balloon and attached experiment circled Earth's South Pole continuously gathering high-resolution data on the CMB. Indeed, the data was of such high quality that researchers were able to measure the curvature of space with it (Fig. 4.8). This important result was made possible by analyzing the spatial distribution and scale of the temperature fluctuations in the CMB. By comparing the observed fluctuation patterns to those expected for closed, open, and flat universes the Boomerang consortium were able to show that the observations are best described when space is Euclidean – that is, in a universe in which $\Omega \approx 1$, if not $\Omega = 1$ exactly.

Fig. 4.8 Small-scale temperature fluctuations in the CMB as revealed by the Boomerang experiment. The distribution of cooler regions (as shown in the *lower three panels*) is consistent with the geometry expected of a flat universe with Ω very close to unity. (Image courtesy of the International Boomerang consortium)

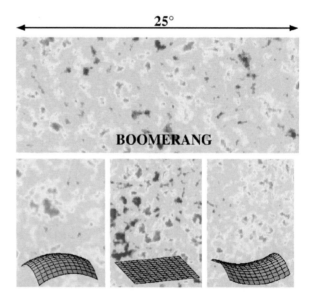

The Boomerang results indicated that the theoreticians, with their favored value of $\Omega = 1$, were possibly right all along. The problem now was to reconcile this result with the other observations. As we have already seen, there is not enough baryonic matter in the universe to bring Ω even close to unity, so there must be other contributing factors. In the next two chapters we shall see that the existence of both dark matter and dark energy are required to bring the observed matter density of the universe up to equality with the critical density, thus satisfying the $\Omega = 1$ constraint set by the CMB temperature fluctuations.

Primordial Nucleosynthesis

After the existence of the CMB, perhaps the next clearest indicator that the universe in which we live must have undergone an extremely hot Big Bang initial phase is the observation that the universe is mostly composed of just hydrogen and helium. Clearly there are other elements – the carbon, nitrogen, and oxygen that is essential for our existence, for example – but these elements are not primordial; rather they have all been produced by multiple past generations of massive stars (as described in Chapter 1). Initially, the universe was almost a pure hydrogen and helium mix, with hydrogen accounting for 75% of the mass and helium the remaining 25%. Importantly, however, trace amounts of deuterium, helium-3, lithium, beryllium, and boron were also produced in the first few minutes after the onset of the Big Bang. Incredibly, by observing the abundance of these trace elements today, the conditions that existed when the universe was just a few minutes old can be accurately deduced.

As soon as the Big Bang occurred the universe started to expand and cool. Initially the temperature was far too high for matter of any kind to exist for very long. The universe consisted entirely of radiation. Remarkably, after about 10 ns of expansion the temperature of the universe dropped to about 10^{14} K, making it cool enough for protons and neutrons to begin forming. This initial creation process continued until the temperature dropped to below 100 MeV (about 10^{12} K), corresponding to the rest mass energy of the proton. The universe is now about 0.0001 s old. Below 100 MeV protons and neutrons will not form out of the radiation field. The basic building blocks of matter were thus assembled, and the next step was to construct larger and larger nuclei – if possible.

Once neutrons had formed within the primordial mix the clock for creating atomic nuclei began its countdown – the race to create matter beyond hydrogen had begun. Unlike protons, which current experimentation shows are stable for all time, a free neutron will undergo beta decay into a proton plus an electron and antineutrino after about 15 min. The only way to avoid this prompt decay is to incorporate the neutron into an atomic nucleus.

Primordial nucleosynthesis is therefore highly dependent upon the expansion rate of the universe and the baryonic matter density. If the former is too fast and the latter too low within the first few minutes of the universe coming into existence, then its composition would be very different from that which we see today. The process of neutron capture into nuclei began once the temperature dropped below about 2.2 MeV, which occurred about 1 s after the Big Bang.

At this stage deuterons, a nucleus composed of a proton and a neutron, began to form. For the next 5 min or so the primordial alchemy continued with the formation

Fig. 4.9 The Big Bang primordial nucleosynthesis of light elements. If the baryon (ordinary matter) density is too small, then there is too much lithium and too little helium-4 compared to the observations. If the baryon density is too high, then there is less deuterium than helium-3, and too much lithium when compared against the observations. The models and observation of light element abundances agree (*thin vertical line*) when the matter density is about 3×10^{-10} that of the photon density, indicating that $\Omega_B = 0.04$. (Image courtesy of NASA)

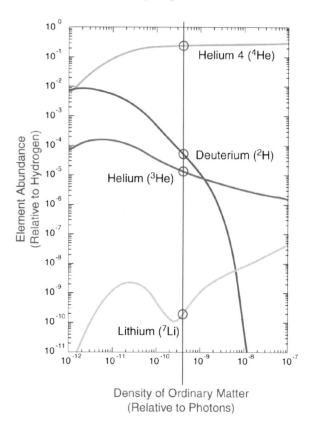

of mostly helium and smaller amounts of lithium, beryllium, and boron. Through constructing detailed nuclear reaction networks physicists have been able to model the expected abundances of the light elements in the early universe as a function of the baryonic matter density (Fig. 4.9). By comparing such model outcomes against the observed abundances for the light elements it has been possible to constrain the baryonic matter density to be about 4% of the total matter density of the universe. Astronomical observations also reveal that most of the baryonic matter in the universe is actually in the form of a diffuse intergalactic gas rather than belonging to higher density stars (Fig. 4.10).

The success in understanding the primordial abundance of light elements and the origin of the microwave background are the two great triumphs of Big Bang cosmological theory. There is indeed no escaping the fact that our universe must have begun in a hot, high density state. There is more, much more, however, that has yet to be fully understood.

On the one hand, observations of the microwave background indicate that we live in a flat (or Euclidean) universe, with the matter density corresponding to the critical value (i.e., $\Omega = 1$), and yet the primordial nucleosynthesis calculations, which match

Fig. 4.10 Chandra spacecraft X-ray image (*left*) of the cluster Abell 2199. The strong X-ray emission indicates that the intracluster medium is very hot, with a temperature of millions of degrees. A detailed comparison of the Chandra X-ray data with the optical data (*image to the right*) indicates that most of the cluster's baryonic mass is in the intracluster gas rather than in the visible stars. (Image courtesy of NASA)

beautifully with the observations, indicate that ordinary matter in the form of visible stars and the gas within the interstellar and intergalactic mediums can only account for 4% of the critical density. The implications are clear; there must be more matter in the universe than literally meets the eye. But where is this apparently missing mass, and what type of matter is it? These questions will be discussed in Chapter 5 and 6. For the moment, however, let us explore the Big Bang in a little more detail, and ask what was possibly happening prior to the formation of the baryons.

Inflation, Flatness, Horizons, and a Free Lunch

We shall never be able to visually observe the very earliest moments of the Big Bang, no matter how large a telescope the astronomers of the future might find funding for. The time of recombination shrouds all previous history in an optical fog of diffuse light. This does not mean, of course, that there aren't observations that can be used to set stringent limits on the types of early universe models that the theoreticians can construct.

Figure 4.11 shows a schematic timeline for the size and evolutionary history of the universe. We have already considered the domain of the "Big Freeze Out," when baryons first formed and primordial nucleosynthesis fixed the chemical composition of the early universe. The time of recombination and last scattering domain "Parting Company" has also been described. The domains corresponding to "First Galaxies"

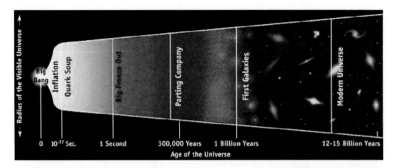

Fig. 4.11 Timeline for the evolution of the universe. The various domains are labeled according to the dominant structures present or processes operating as the universe aged. (Image courtesy of ESA)

and the modern universe will be the topics of Chapter 5 and 6. Our task here is to consider what was going on in the universe prior to the time that it was 1 s old.

At first thought one might be fooled into thinking that not much can happen in 1 s. We might click our fingers twice in such a time interval, and even the fastest piloted aircraft in the world (NASA's experimental X-15 rocket-powered aircraft) could travel but a mere 2 km. In the subatomic world, however, a near infinity of interactions, exchanges, and decays can take place within a second, and for the early universe the first second was enough time for its size to increase by an incredible factor of about 10^{70}.

An absolutely humongous increase in the size of the universe began shortly after it first came into existence – indeed, the process began when the universe was just a miniscule 10^{-35}-th of a second old. This process of rapid and expansive inflation was first described by Alan Guth (Massachusetts Institute of Technology) in the early 1980s, and it is based upon the idea that the universe underwent a dramatic, energy-generating phase transition during the time interval encompassing 10^{-35}–10^{-33} of the very first second. The detailed physics of the phase transition need not concern us here; the key point is that the universe expanded dramatically. The existence of an inflation era, it turns out, is very important for explaining two significant cosmological problems – the flatness problem and the horizon problem. The first of these problems relates to the question which asks why Ω is apparently equal to unity at the present epoch, while the second problem relates to the incredibly uniform temperature of the microwave background.

The flatness problem addresses what is inherently a surprising selection effect. That we should find ourselves in a universe in which $\Omega = 1$ is, upon examination, rather odd. Peter Coles (Cardiff University, Wales) has presented a nice analogy to describe the situation in which cosmologists find themselves. Imagine that we are told that behind a door that we can open at any time we choose is an acrobat on a tightrope. Now, there are two most likely states that we might expect to find the acrobat in when we eventually open the door: he will either be on the tightrope, or he will be on the floor (having fallen off the tightrope). We would indeed be highly surprised if, exactly at the time we eventually opened the door, we caught the acrobat in the process of falling off the tightrope.

The fact that we live in a universe in which $\Omega = 1$ is akin to opening the door and finding the acrobat neither on the tightrope nor on the floor but in mid-flight. We can think of the $\Omega > 1$ and the $\Omega < 1$ states as corresponding to the acrobat being either on the tightrope or on the floor. Indeed of all the possible numbers and fractions that exist between zero and infinity, selecting out the number one is highly improbable, and yet this is exactly what we find the universe has apparently done. This is a startling result.

To elucidate the problem further we can look to the Friedmann solution sets to Einstein's equations, and from these it can be shown that Ω must evolve as the universe ages. A glance back at Fig. 4.3 reminds us that when the universe was very young the exact value of Ω was not greatly important (it could be greater than, less than, or equal to unity) since at that time all the possible expansion paths are indistinguishable. The problem, however, is that even very small differences in Ω away from unity will grow very rapidly as the universe ages, and we should see the consequences of this difference in the universe around us.

For example, if $\Omega = 0.9$ held just 1 s after the Big Bang, then its value now, some 14 billion years on, would be about 10^{-14} (i.e., essentially zero). If, on the other hand, $\Omega = 1.1$ held just 1 s after the Big Bang, then the universe would have collapsed and undergone a Big Crunch after just 45 s – though perhaps we should say Little Crunch, given the circumstances. For what must be very special reasons, therefore, we apparently live in a universe that has been fine tuned to a very high degree of accuracy. There is nothing in science that says that surprising circumstances can't come about, but their appearance does suggest that there is some additional underlying principle that has yet to be understood.

Inflation, it turns out, is the key to understanding the flatness problem. Indeed, the dramatic burst of growth during the inflation era will smooth out any initial geometrical wrinkles that the universe might have had. If we go back to our balloon analogy, introduced earlier in relation to Hubble's law, then, when it is very small (i.e., at a time corresponding to the very early universe) its surface is highly curved and wrinkled. If we now imagine inflating the balloon to say the size of Earth, then its surface will become so stretched out that it will appear flat in a region limited to, say, a few meters or possibly kilometers in extent. Accordingly, any one such flat region on the greatly inflated balloon's surface is an analog of our observable universe (see Fig. 4.12).

Solving the flatness problem in terms of a period of early rapid inflation also provides us with an answer to the horizon problem. If we take another look at Fig. 4.6 the temperature of the cosmic microwave background is a near perfect fit to a blackbody curve, with a temperature of 2.725 K. This is totally remarkable, and it tells us that the Big Bang must have occurred under conditions of essentially perfect thermal equilibrium, with the temperature being virtually the same everywhere.

There is a problem, however, with this argument, and the problem like so many of the issues in our everyday lives is one of communication. In anthropomorphic terms, we might say that to be in thermal equilibrium requires that all regions of the

Fig. 4.12 Inflation drives the universe, irrespective of its initial curvature, into a flat, Euclidean state. The analogy illustrated here is that of an initially warped and wrinkled balloon being inflated to a very large size. What eventually becomes our observable universe will, after inflation, be geometrically flat

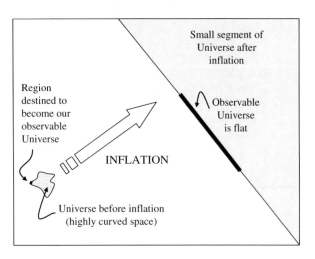

universe have exchanged information about their temperature and accordingly the hot and cold regions have come to an agreement on what the common temperature should be. That is, hot areas essentially redistribute their heat to cooler regions until a constant, uniform temperature is achieved. In essence this is just straightforward thermodynamics, and a process that is easily demonstrated within any coffee shop. The problem for the universe, however, is that it is simply too large for the information exchange necessary for thermodynamic equilibrium to have come about, to have taken place. Even with the information being transmitted at the speed of light, the fastest speed with which any information can be transmitted, two regions of the microwave background separated by the diameter of the observable universe (a distance of about 25 billion light years) could never have communicated their existence, let alone come into thermal equilibrium. This is the horizon problem.

Inflation solves the information exchange issue that lies at the heart of the horizon problem by positing that what is now our observable universe was initially just a miniscule region of the original Big Bang creation. In the inflation model our observable universe began as a region perhaps just 10^{-27} m across. In a domain this incredibly small radiation can rapidly come into local thermal equilibrium, and this feature becomes locked in after the epoch of massive expansion brought about by inflation.

Although the temperature of the microwave background is incredibly uniform, it is not perfectly so. A look at Figs. 4.8 and 4.9 reminds us that there are temperature variations on a level of order 10^{-5} K. What, we might ask, caused these temperature variations to come about? The answer, once again, is inflation, or more to the point the size-amplifying properties of inflation. Incredible as it may seem, the fluctuations or ripples observed in the cosmic microwave background today are derived from quantum fluctuations produced when the universe was about 10^{-35} s old.

Normally, quantum fluctuations operate on the smallest of atomic scales and come about in accord with Heisenberg's uncertainty principle, which allows small changes in the energy to spontaneously occur for very short intervals of time. Space, even what is apparently empty space, is, at the quantum level, a seething cauldron of particles and their antiparticles popping into and then out of existence – the energy required to create the particles being briefly borrowed and then returned as they are annihilated. The universe is literally founded upon an ever-shifting froth of virtual particles that disappear almost as soon as they form. During the inflation epoch, however, these barely existing quantum fluctuations are stretched out and amplified to gargantuan size. Incredibly, the temperature variations that we now see in the CMB are directly related to these stretched out primordial quantum fluctuations. In the macrocosm is the microcosm.

Figure 4.13 shows the observed power spectrum of temperature fluctuations against angular size in the cosmic microwave background. This diagram reveals how the temperature of the microwave background varies from point to point across the sky against angular frequency (this last term measures the number of fluctuation cycles that are present around the entire sky). The first peak at an angular scale of just over 1° depends upon the total matter-energy density of the universe at the time of recombination (when photons were no longer efficiently scattered by electrons), and it is this feature that essentially tells us that we live in a universe in which the matter density is exactly equal to the critical density.

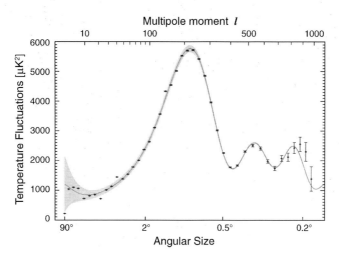

Fig. 4.13 Small scale angular variations in the temperature fluctuations of the cosmic microwave background radiation. The ripples essentially provide us with a direct experimental probe of the conditions that prevailed in the universe when it was a sprightly 10^{-35} s old. (Image courtesy of NASA)

The other, lesser peaks, at smaller angular scales that can be seen in Fig. 4.13, are related to density fluctuations caused by sound waves and provide a constraint on the value of Ω_B. These features also relate to interactions that are set up between the slightly clumped baryons and photons in the primordial Big Bang plasma, and most importantly they are diagnostic features that in principle tell us what the matter content of the universe is, what value applies to Hubble's constant, whether there is dark energy (the subject of Chapter 6), and even if there really was an inflation epoch or not. Figure 4.13 is literally packed full of cosmological goodies, and researchers will be improving upon its resolution and analyzing its contents for many years to come.

That the universe is flat or incredibly close to being flat appears to be an inescapable fact of modern-day observational cosmology. It is still not fully clear, however, that Ω is exactly equal to unity, but many theoreticians feel the strong compulsion to believe that it must be so. Indeed, as Alan Guth and many others have emphasized over the years, a flat universe is aesthetically beautiful, and it also gives the lie to the statement that there is no such thing as a free lunch. The latter point comes about since in a flat universe, and in no other, the total energy content is exactly zero. So, while there is lots of motion in the universe, stars and galaxies all moving in multitudinous abandon, there is a fine balance between the total kinetic energy (the energy of motion) and the gravitational energy of all the objects being carried along with the universal expansion.

An analogy of this situation can be envisioned by considering the act of throwing a ball. We all know that if we throw a tennis ball directly upwards that it gradually slows down with increasing height, coming eventually to a standstill and then falling back to Earth. In this case the initial kinetic energy of motion was overcome by the potential energy due to Earth's gravitational field. This situation is comparable to that of a closed universe, with $\Omega > 1$, which will eventually collapse towards a Big Crunch. If we now imagine throwing the ball upwards very much harder, perhaps making use of a powerful canon, then the ball will rise and rise, its velocity hardly slowing, and it will never come back down again. In this case the kinetic energy of motion always exceeds that of the potential energy due to Earth's gravitational field. This situation is analogous to an open universe in which $\Omega < 1$.

Situated in between these two conditions of always falling back after a finite time interval and never stopping, we have the case where the kinetic and gravitational potential energy are exactly equal and opposite – that is, the total energy is zero. Under these conditions if we throw the ball upwards with just the right speed, the so-called escape speed, then it will rise and rise, slowing down continuously, but never coming to a complete stop until an infinitely long interval of time has elapsed, and the ball is infinitely far away from Earth. Skipping over the details of what exactly we mean by infinity, other than referring to numbers that are incredibly large, we find that a zero energy universe will expand forever. Remarkably, therefore, we can think of a flat universe, in which the matter density is exactly equal to the critical density, as having arisen out of nothing and which will run forever on borrowed time. One is inclined to say, therefore, that an inflated flat universe, on physical as well as economical grounds, is one heck of a good deal.

The Quark–Gluon Plasma

Take another look at Fig. 4.11. According to this timeline diagram, the inflation epoch ended when the universe was just 10^{-32} of a second old. At this stage the

Fig. 4.14 Road map to the theory of everything, illustrating how in the very early universe all of the four fundamental forces combine to act as one grand unified force. The forces separated out in three phases; gravity first, after the first 10^{-43} s (the Planck time), ending the quantum gravity (*speckled region*) era. Later the strong nuclear force and the electroweak force (*dark grey region*) separated out, and finally after about 10^{-12} s the weak nuclear and the electromagnetic force separated out from the electroweak force. The *boxes at the top* of the diagram illustrate the connections between the fundamental forces and their domains of physical study at the present time

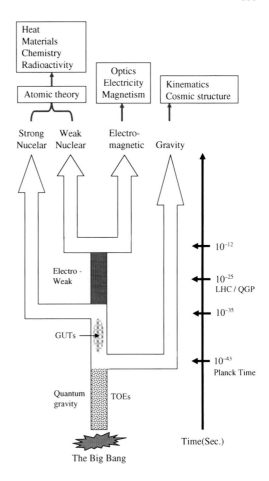

temperature of the universe, now just a few centimeters across, was about 10^{27} K. At these colossal energies we are still far removed from the realm at which the LHC will operate, but we are very much in the domain where theoretical physicists predict that the electroweak and strong nuclear forces begin to become distinct (see Fig. 4.14 and recall Chapter 3). By the time the universe is 10^{-12} s old, however, its temperature will have dropped to about 10^{15} degrees, and the electroweak force will branch into the distinct weak nuclear force and the electromagnetic force.

At this stage we are well within the realm of terrestrial collider experiments. Indeed, the LHC, when operating at full power, will be able to momentarily recreate the conditions that prevailed when the universe was a minuscule 10^{-25} s old with a characteristic temperature of 10^{17} K. At this energy range the LHC will be able to explore the quark-gluon plasma era of the early universe. It is this phase that prevails until the so-called "big freeze out" when the top and bottom quarks (recall Fig. 3.1) begin to combine and produce protons and neutrons.

As we noted in Chapter 3 the Standard Model dictates that at the low temperatures that prevail in the present-day universe all quarks and gluons are trapped within the confines of protons and neutrons; they are lifers, with no chance of parole, imprisoned within myriad baryonic jails. When the universe was less than of order 10^{-12} s old, however, the quarks and gluons had a period of freedom, or deconfinement, allowing them to move relatively freely in a neutrally charged particulate soup that has been named the quark-gluon plasma (QGP).

The LHC will directly study the QGP by crashing beams of lead (Pb) ions together at energies approaching 1,150 TeV. Such experiments will enact the ultimate smash and grab jailbreak for the quarks and gluons ordinarily held captive within their locked-tight baryon cages. The key QGP experiments will be conducted with the ALICE detector (described in the next paragraph) along with additional experiments on the ATLAS and CMS detectors. It is remarkable, indeed, even incredible to think that LHC researchers will be able to probe a state of matter that hasn't existed for the past 14 billion years, and that when the QGP last existed it endured for only about one 10 millionth part of the very first second of the universe blinking into existence.

As with all collider experiments the idea behind ALICE (A Large Ion Collider Experiment) is to induce collisions between nuclei and to then study the particulate shrapnel produced. The LHC experiments will use ionized lead atoms (Pb) in the colliding beams, rather than protons, since this will ensure that there are many potential target nuclei. Each lead atom nucleus contains, in fact, 82 protons and 125 neutrons. Provided that the energy is high enough, and it certainly should be in the LHC, then Pb-Pb collisions will generate a vast, expanding fireball of interacting, deconfined quarks and gluons. As the GQP fireball expands and cools, however, rehadronization will take place, with the quarks and gluons once again being swept up into their atomic jail cells (Fig. 4.15).

Fig. 4.15 Simulation of the fireball spray of charged particles produced during a Pb–Pb collision within ALICE. The job of the various instruments nested within the detector is to identify and track the time of flight and spatial motions of these numerous particles. (Image courtesy of CERN)

The existence of a QGP phase of matter is one of the key predictions of quantum chromodynamics (QCD), and the researchers at CERN intend to use ALICE to measure the production rate of quarkonium – a meson composed of a quark and its own antiquark. Initial low energy collider experiments carried out at Brookhaven's RHIC and CERN's SPS indicate that the quarkonium production rate is sensitive to the collision dynamics that prevail within the collider, with the charmonium (the meson composed of a charm quark and anticharm quark pair) production rate, in particular, being suppressed once a QGP has formed. The ALICE forward scatter muon spectrometer (see below) will directly study the charmonium and bottomium (the meson composed of a bottom and antibottom quark pair) production rates via their decay into muons through the $\mu^+\mu^-$ channel. The QCD predictions indicate that at the high collision energies that will prevail at the LHC it is likely that charmonium production should be enhanced, rather than suppressed, since many more deconfined charm and anticharm quarks should be generated per collision. The greater production numbers per collision at the LHC should enhance the chances of a deconfined charm quark meeting a prospective deconfined anticharm quark, to produce a charmonium meson.

ALICE: In Experimental Wonderland

There is nothing shy about the scale of the ALICE; it is huge (Fig. 4.16). Its approximate dimensions read as 26 m long, 16 m high, and 16 m wide, and its total weight tips the scale at about 10,000 tons. Within its girth resides the world's largest solenoid magnet, capable of generating a field of strength 0.7 T. For comparison, this is about 15,000 times the field strength of Earth's magnetic field. This vast machine has been designed and refined by a vast international collaboration consisting of

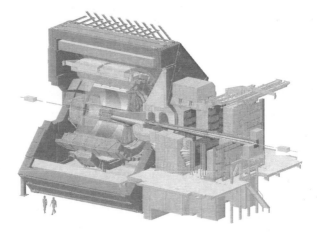

Fig. 4.16 Schematic cut-away diagram of the ALICE detector. Note the to-scale humans shown towards the *lower left hand corner*. (Image courtesy of CERN)

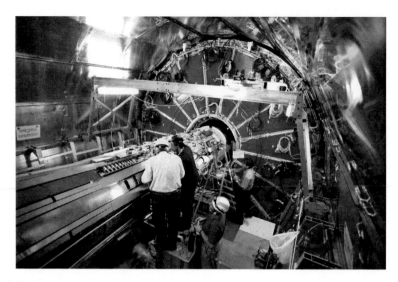

Fig. 4.17 Technicians at work on the inner tracking system (ITS) located at the very core of ALICE. (Image courtesy of CERN)

over 1,000 researchers, from 109 institutes in 31 countries, and it has taken nearly 5 years to build (Fig. 4.17) following a 10-year design phase starting in the early 1990s.

ALICE is an incredibly complex machine, which is fitting because the task that has been set for it is gargantuan: to identify and then measure both the time and spatial evolution of the 10,000 or more charged particles that will be produced during a Pb–Pb collision (Fig. 4.15). To achieve this end ALICE is really a nesting of different detectors working in tandem and simultaneously contributing to the overall particle-tracking procedure.

At the heart of ALICE sits the inner tracking system (ITS) – see Fig. 4.16. This cylindrical detector hugs the 1-mm diameter beam pipe, and its six layers of silicon detectors will track the short-lived particles that decay before traveling more than a few millimeters away from the interaction point. Moving further outwards from the central beam pipe the ITS is surrounded by the time projection chamber (TPC). This detector is the largest of its kind ever built, and its more than 560,000 readout channels will provide a near continuous three-dimensional tracking of the charged particles at distances of between 80 and 250 cm away from the interaction point.

Moving still further outwards the next instrument encountered is the transition radiation detector (TRD) that, with over 1 million readout channels, will identify electrons. Wrapped around the TRD is a time of flight (TOF) detector that will measure the total travel times of the particles from the interaction point to the detector edge. The TOF system works to an incredible resolution of about 100 ps. The outer wrapping of ALICE is completed by a set of detectors, one of which will identify and measure photon energies, thus allowing the temperature of the fireball to be determined, while the other will measure the properties of high momentum hadrons.

Fig. 4.18 The dipolar magnet assembly (*right*) and one of the muon spectrometer filter walls (*left*). (Image courtesy of CERN)

The radial extent of instruments is now complete, but ALICE has still more measuring devices that it can call upon. In one forward direction (to the left in Fig. 4.15) sits a detector to measure the spray of photons that will be produced during a collision. Likewise, in the opposite forward direction (to the right in Fig. 4.15) is a large muon spectrometer (Fig. 4.18). This system has been specifically designed to study the spectrum of heavy quarkonium states (e.g., charmonium and bottomium) via their decay into the $\mu^+\mu^-$ channel – recall it is the production rate of heavy quarkonium states that should change once a QGP has formed.

The many instruments nested within ALICE will generate a considerable amount of data during the 1 month/year that the LHC will be in heavy-ion Pb operation mode. Although only those events producing results of specific interest are going to be recorded it is estimated that of order a 10^{15} bytes of data will be generated per observing cycle – a staggering amount of data. Given the high data generation rates it is expected that ALICE will begin to yield new data on the QGP after just one operational cycle.

Matter/Antimatter: It Matters!

Why is there something rather than nothing? This is a question that the great philosopher and theologian Thomas Aquinas asked in the thirteenth century. It is a very good question, and it is one that physicists and cosmologists are only now just coming to terms with.

Indeed, the fact that we live in a matter-dominated universe is a major unresolved problem. At issue is the inherent fairness of particle creation and the expectation that from a given energy input an equal number of particles and antiparticles should be created. In other words, the expectation is that the Big Bang should produce an equal number of particles (constituting matter) and anti-particles (constituting antimatter), and yet here is the rub. We live in a matter-dominated cosmos. Where have all the antimatter particles gone?

In theory there is absolutely no reason why antimatter can't clump together to form antistars and antigalaxies, but there is no strong observational evidence to support the existence of such structures within the observable universe. Likewise, whenever matter and antimatter particles encounter one another there should be a great burst of energy generated as they annihilate (as illustrated in Fig. 3.8). If an electron, for example, encounters its anti-particle equivalent, the position, then the annihilation energy liberated, will be equivalent to $e^- + e^+ \Rightarrow E = 2m_e c^2$ (which amounts to 1.6×10^{-13} J ≈ 1 MeV of energy). Once again, however, no such transient bursts of annihilation radiation are observed. So, we ask again, where has all the antimatter gone?

The best answer at the present time is that we simply don't know for sure, but it is one of the fundamental questions that the LHC experimenters will try to address. Indeed, the LHCb detector (to be described shortly) has been specifically built to study the key physical issue encapsulated in the phenomenon known as CP violation.

To end up with a matter-dominated universe it has been calculated that only one matter particle need survive in 30 million annihilation events. This is a remarkably fine-tuned situation that most physicists suspect didn't just come about by pure chance. This being said, however, a case (not a popular one, it should be added) can be made for the argument that the universe simply formed in a matter-dominated fashion, and that the whole matter-antimatter problem is a non-issue. Most of the current research relating to the matter-antimatter problem, however, centers on the idea that there must be a subtle difference between the way in which particles and anti-particles behaved in the primordial fireball of the Big Bang. The first hint that this might have been so was observed experimentally in the mid-1960s as part of a study related to the decay of kaons (a form of meson containing either a strange or an antistrange quark). Indeed, the 1980 Nobel Prize for Physics was awarded to James Cronin and Val Logsdon Vitch for their discovery of what is known as CP violation – the process that accounts for the fact that nature does not always treat quarks and antiquarks in the same manner under the same conditions.

CP violation relates to the idea that the laws of physics should give the same result if a particle is transformed into its mirror image antiparticle. Theory indicates and experiments also show that under strong and electromagnetic interactions CP-symmetry holds true, but, and here is the important point, this symmetry can be violated under certain kinds of weak decay – as first shown by Cronin and Vitch. At best it can be said that CP violation indicates that, as it currently stands, the Standard Model is incomplete, and this, of course, is exactly why the LHC has been built.

The fact that QCD preserves CP symmetry is a mystery in its own right (called the strong CP problem). One attempt to solve this issue has been to introduce a new host of charge-less, low mass, weakly interacting particles called axions. Such particles, it has been suggested, should have been copiously produced in the Big Bang and may presently pervade the universe, perhaps as dark matter. According to theory axions are predicted to change into photons when they pass through a strong magnetic field, which opens the potential doorway for observational experimentation. No such experiments are planned to be conducted at the LHC, however.

Getting to the Bottom of Things

The LHCb detector is some 21 m long, 10 m high, and 13 m wide; weighs in at 4,500 tons, and is composed of a linear series of detectors that will study the forward-moving spray of particles and their decay products generated during a proton–proton collision (Fig. 4.19). As with all the CERN instruments, the LHCb has been designed and built by a large consortium of researchers. The primary purpose, and indeed guiding design principle, is to study to production and decay of bottom quarks. Specifically it is the decay of B-mesons, mesons containing the bottom and antibottom quarks, that the LHCb researchers wish to look at.

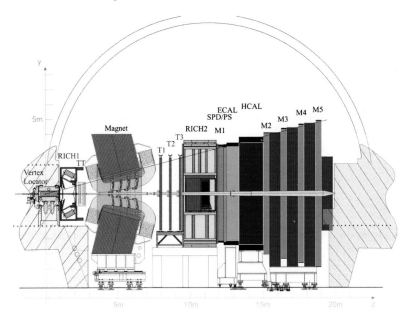

Fig. 4.19 Schematic layout of the LHCb detector system. (Image courtesy of CERN)

Following a P–P collision at the entrance aperture of the LHCb complex of instruments, the B-mesons will move downrange, staying close to the beam pipe rather than spraying in all directions. The first instrument within the LHCb is the vertex locator, which will pick out the B-mesons from the multitude of other collision products and channel them into the first Ring Imaging Cherenkov detector (RICH). This

Fig. 4.20 The massive LHCb dipole magnet consists of two 27-ton coils mounted inside of a giant 1,450-ton steel cage. Each coil is 7.50 m long, 4.6 m wide, and 2.5 m high and is made of nearly 3 km of aluminum cable. (Image courtesy of CERN)

devices looks for the light pulses that will be produced when high-speed particles stream through the detector's silica aerogel and perfluorobutane gas-filled interior. (The light pulses are caused by the production of what is called Cerenkov radiation, emitted when a particle moves with a speed greater than that of light in the specific medium.) By monitoring these light pulses the velocities of the various particles can be determined.

After passing through RICH1 (Fig. 4.19), the particles will pass through a large dipole magnet (Fig. 4.20). This massive magnet will accordingly deflect the paths of any charged particles, with positive and negatively charged particles moving in opposite directions. By examining the curvature of the particle paths their momentum can be measured and their identity can also be established.

Upon leaving the magnetic cavity the particles then encounter a series of tracking systems (T1, T2, and T3 – Fig. 4.19) that will enable the trajectory of each particle to be recorded. After passing through a second Cherenkov detector (RICH2) the particles will then encounter a series of calorimeters. The first instrument, the electromagnetic calorimeter (ECAL), will measure the energy of the lighter particles such as the electrons and photons. The second, hadron calorimeter (HCAL) will determine the energies of the heavier particles such as any protons and neutrons. Finally, at the far downrange end of the detector, are a series of muon detectors (M1 to M5). Since muons are present in the final states of many CP-sensitive B-decays the study of these particles will play a vital role in understanding the CP asymmetry that is believed to account for matter dominance in the universe.

When fully operational the LHCb detector will register an estimated 10 million proton collisions each and every second – far too many events to follow individually.

A trigger system has therefore been introduced to record just 1 million of the events for further processing. This initial data grab is fed at a superfast rate of 35 Gbps to a series of some 2,000 computers, which will further refine the analysis to archive the most interesting 2,000 events. Even at this reduced sampling rate the LHCb detector will generate about 250 Gb of data per hour (equivalent to the data contained on about 300 CDs). During the course of a year the system will accrue a data set that would fill the equivalent of 400,000 CDs – a stack comparable in height to the Empire State Building in New York City!

By the time that the universe was just 1 s old it was already matter dominated. The LHCb experiment will begin to explore the reasons for why this (presently paradoxical) condition came about, and eventually it will help researchers answer, as St. Thomas Aquinas so long ago asked, why there is something rather than nothing in the universe.

Chapter 5
Dark Matters

Towards the close of the eighteenth century William Herschel, the celebrated British king's astronomer, set out to count all the stars in the sky. It promised to be an onerous task, but with the help of his ever-faithful sister and scribe Caroline, it was a project that he felt was doable. Working predominantly with a 20-ft reflector (Fig. 5.1), under the less than favorable night skies of Slough in England, the Herschel team collected star gauges from selected regions of the heavens. Herschel counted and Caroline recorded, and from the collected star numbers the first ever map of the large-scale distribution of cosmic matter was constructed.

In order to produce his map, Herschel assumed that all stars were of the same luminosity and that his telescope could reveal stars out to the full extent of their distribution in space. Both of these assumptions are now known to be wrong, but unlike Herschel we have the benefit of hindsight and several centuries' worth of additional study to rely upon. The stellar universe that Herschel's star gauges revealed was disk-like in shape, with the Sun located almost exactly at its center. In addition, almost one half of the disk was apparently cleaved open along what is seen as the band of the Milky Way (Fig. 5.2).

In 1789, shortly after seeing the Royal Society publish his star-gauging paper, Herschel oversaw the completion of his 40-ft focal length reflector. This great telescope, although highly unwieldy in operation, was Herschel's crowning triumph, and it was the marvel of the entire world. Sadly for Herschel, however, he soon discovered that with his new larger telescope he could see many more stars than he had with his 20-ft reflector. The distribution of stars, he realized, must stretch much further into space than he had deduced from his earlier star gauges.

It was also towards the close of the eighteenth century that Herschel realized that stars could form relatively close, gravitationally bound binary pairs, and further, he could observe that in some case the two stars had different brightnesses. This latter observation revealed that the stars in general must have a wide range in intrinsic luminosity – not all stars, it appeared, were created equal.

In spite of Herschel's pioneering efforts, the dawn of the nineteenth century saw all his earlier deductions cast into doubt. Not only was his distance estimation method incorrect (stars are not all cast from the same mold), but the true extent of the stellar system could not be reached with any certainty. Larger telescopes simply

M. Beech, *The Large Hadron Collider*, DOI 10.1007/978-1-4419-5668-2_5,
© Springer Science+Business Media, LLC 2010

Fig. 5.1 The 20-ft reflector that Herschel used in his star gauges. The telescope was completed in 1783 and is named according to the focal length of its 18-inch diameter mirror – a mirror that Herschel had carefully cast and polished himself

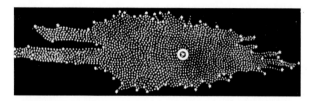

Fig. 5.2 Herschel's stellar universe. This image shows a cross-section through the stellar disk described by Herschel in his 1785 research paper published by the Royal Society of England. The "cleave" in the disk is visible to the left where the star distribution apparently splits into two channels on either side of the Milky Way

revealed more and more faint stars. Not only this, however. Astronomers were also finding more and more oddly shaped nebulous systems. Indeed, it was to be these diffuse clouds of nebulosity, which refused to be resolved into stars, that ultimately allowed astronomers to determine the truly gigantic scale of the universe. Herschel once famously wrote that, "I have looked further into space than any human being did before me." Unfortunately for Herschel, and all other astronomers before and since, one might follow such a statement with the rejoinder, "and what you saw through your telescope is not necessarily all that there is to be seen."

Interstellar Matters

At sea level, the air we breathe contains about 3×10^{25} molecules of matter per cubic meter. This number decreases with increasing altitude, dropping by a factor of 2 billion by the time a height of 150 km has been reached. Moving further outwards from Earth through the Solar System, past the Oort Cloud of cometary nuclei and on into interstellar space the number of atoms per cubic meter becomes smaller and smaller, but it never quite drops to zero.

The average number of atoms in interstellar space is of order 1 million atoms per meter cubed, although the number can vary anywhere from 10,000 to over a billion atoms per cubic meter. Even at its most extreme, however, the density of the interstellar medium (ISM) is still more than 1,000 times smaller than the ultra-high vacuum conditions that will operate within the LHC beam line. In terrestrial laboratories, an ultra-high vacuum contains of order 300 billion molecules/m^3.

Space is not, in spite of first appearances, empty. Rather, it is pervaded by a diffuse ISM composed of a complex intermingled, intermixed, and clumpy distribution of gas and dust. Hydrogen and helium atoms are by far the most common elements within the ISM, with only about 1% of the ISM's total mass being in the form of minuscule, solid silicate and carbon dust grains (Fig. 5.3). The hydrogen and helium is predominantly primordial and was produced during the time of primordial nucleosynthesis, just after the Big Bang.

Eventually, in the far distant future, all of the hydrogen and helium in the ISM will be transformed into stars and thereafter synthesized into atoms of carbon and atomic elements beyond. The dust component of the ISM, however, will continue to

Fig. 5.3 The dense interstellar dust cloud Barnard-68 looks to the eye like a giant hole in space. Located some 500 light years from Earth the nebula is about one half of a light year across and contains a sufficiently large number of dust grains to fully absorb and scatter from our view any starlight entering the cloud from its far side. (Image courtesy of ESO)

grow as our galaxy ages, since it is produced within the outer envelopes of old, low-mass stars and within the nebulae surrounding supernova disruption events. Our view of the heavens, at least within the disk of the galaxy, is destined to become more and more obscured.

When conducting his pioneering star gauges William Herschel had no idea that the ISM existed. It was for this reason that Herschel believed he should be able to see to the very edge of the stellar system with his various telescopes. It was the interstellar dust, the minor component of the ISM, that defeated Herschel in his surveys. Regions of the sky such as Barnard 68 (Fig. 5.3) were interpreted by Herschel (not unreasonably at that time) as being devoid of stars and therefore regions of totally empty space – the literal edge of the observable universe. In fact Barnard 68 is a region in which two interstellar gas clouds are colliding. The larger cloud has about twice the mass of our Sun, while the smaller cloud (visible in the lower left of the image) contains just a few tenths of a solar mass of material. Recent computer simulations by Joao Alves (Director, Carlar Alto Observatory) and Andrew Burkert (University of Munich) suggest that a star (or stars) should form in Barnard 68 within the next 200,000 years. The lesson for us here, of course, is that just because we don't see something happening with our eyes or with optical telescopes doesn't mean that nothing is going on.

By chance, it turns out that the typical dimensions of interstellar dust grains are about the same as the wavelength of blue light: $\lambda_{blue} \sim 3 \times 10^{-7}$-m. What this means is that blue light is preferentially scattered by the dust grains, while longer wavelength red light is less dramatically affected. As starlight propagates through the ISM, therefore, the blue light component is repeatedly scattered and the light becomes more and more red light dominated. Stars are dimmed and reddened by the ISM, and accordingly this limits how far we can see into space. Most of the interstellar dust is concentrated within the disk of our galaxy, and this is where the reddening effect is at its most extreme. Indeed, at optical wavelengths we can see only a few thousand light years into the disk. The galactic center, for example, is completely cut off from visual observations because of the reddening and absorption due to interstellar dust.

Let us do a back-of-the-envelope calculation. Present day studies indicate that the interstellar medium accounts for about 10% of the mass of the visible matter in our galaxy, and that about 1% of this mass is in the form of interstellar dust. These combined percentage estimates suggest that there is something like 30 million solar masses worth of interstellar dust within our galaxy – a quantity of order 6×10^{37} kg. If we take a typical dust grain to be about 10^{-7} m across (i.e., about the wavelength of blue light), then each grain will have a mass of about 10^{-17}-kg, and hence, the total number of dust grains in our Milky Way galaxy comes out to a grand total of about 8×10^{54} particles.

If we now imagine packing all of these grains into one large cube, what is the size of the cube? Taking our imagined cosmic dust cube to be of side length D, then its volume will be D^3. This volume must contain the total volume of all of the interstellar dust grains, which will be of order the total number of grains multiplied by the individual grain volume. Hence, we have the result that $D^3 \approx 8 \times 10^{54} \times (10^{-7})^3 = 8 \times 10^{33}$ m^3. In other words, the side of our imagined cosmic dust cube is about 200 million kilometers – a distance equal to about four thirds of an astronomical unit (AU). Again, in other words, if all of the interstellar dust within our galaxy could be gathered together into a vast cube, the grains all packed together, one against each other, then the cube would have a side length a little bit larger than Earth's orbital radius around the Sun. Compared to the size of our galaxy (which is an approximate disk with a radius of about 3 billion AU and a thickness of 65 million AU), there isn't really very much dust in it at all. Incredibly, just 1,000 dust grains/km^3 of space are enough to completely obscure our view (at optical wavelengths) of the galactic center.

With improvements in technology and the introduction of telescopes that work at wavelengths outside that of the visual, astronomers have found that the effects of interstellar dust are wavelength dependent. The shorter the observing wavelength, the greater is the absorption effect, and our galactic view is most obscured. Moving to longer observing wavelengths such as in the infrared and radio parts of the electromagnetic spectrum, however, the effect of the dust obscuration decreases, and more pristine vistas open up to the astronomer's gaze.

Making sense of what we see in the universe, and especially what we see within our Milky Way Galaxy, is a complex, non-trivial problem, and history tells us that we have to be very careful in our interpretations. At any one instant we have but a limited view in space, time, and wavelength region, and in some sense, our attempts at observing the universe are a bit like trying to determine the shape of an elephant by the sense of smell alone in a darkened room.

The medium of astronomical information gathering is the domain of electromagnetic radiation, either through absorption or emission processes. Modern astronomers can now routinely explore the universe by utilizing instruments that cover the entire electromagnetic spectrum, from short wavelength gamma rays to long wavelength radio waves (hopefully, in the not too distant future, gravitational wave detectors will be added to the observational toolkit), but there are still phenomena that cannot, at least at electromagnetic wavelengths, be directly observed. Dark matter is one such phenomenon, and here, to paraphrase Marshall Mcluhan, the medium is the message. Dark matter is not, like Barnard 68 (Fig. 5.3), a dark region of space at one wavelength (i.e., optical) and transparent at another (i.e., radio wavelengths), but it is rather an all-pervading medium that has absolutely no discernable effect on electromagnetic radiation. Unlike the ISM, and the larger-scale intergalactic medium, we presently have no idea what the constituents of dark matter are. Dark matter could be one thing, or as is more likely, it could be many different kinds of particles. Further we have no idea how dark matter interacts with ordinary matter – if indeed, it directly interacts with ordinary matter at all.

Dark matter is indeed the stuff of mystery, magic, and the almost unbelievable. Any yet, we know it exists. This is a truly remarkable state of affairs. Indeed, the evidence for the existence of dark matter is impeccable and totally undeniable. Even though we cannot study dark matter through any direct electromagnetic process we know that it must exist because of its gravitational influence. The existence of dark matter is betrayed, and written large on the sky, by the motions of the stars and the dynamics of individual galaxies within galaxy clusters.

Before we discuss how the uncontroversial evidence for the existence of dark matter has come about, we should first say something about how astronomers discovered our Milky Way Galaxy and deduced the properties of its shape, stellar content, and dynamical structure.

Where Are We?

"That the Milky Way is a most extensive stratum of stars of various sizes admits no longer to the least doubt; and that our sun is actually one of the heavenly bodies belonging to it is as evident." So wrote William Herschel towards the end of his life – and he was absolutely right. It was to be the mid-nineteenth century, however, before astronomers could take such very general sentiments much further. Only in the wake of new photographic surveys and the construction of detailed stellar catalogs that mapped out the locations and brightnesses of hundreds of thousands of stars could a statistical analysis of the stellar system truly begin, and it was soon realized from these that the distribution of stars was not uniform.

Working from the Munich Observatory in the late 1880s, German astronomer Hugo von Seeliger used the famous *Bonner Durchmusterung* star catalog to show that the number of stars increased by about a factor of three per one magnitude decrease in brightness. If space was filled equally with stars (of the same luminosity), then the increase should be by a factor of four per magnitude decrease. Seeliger concluded, therefore, that the density of stars (the number of stars per unit volume of space) must decrease with increasing distance from the Sun. But this result ignored the intrinsic variation in stellar brightness – the so-called luminosity function.

Rather than rely solely upon magnitude data other astronomers began to look at the systematic dynamics of the stars via their measured parallax and proper motion. The parallax data provides a direct measure of the distance of a star from the Sun, while the proper motion data reveals how fast the star is moving through space and in which direction. Jacobus Kapteyn (University of Utrecht) used such data in the early twentieth century to argue that the star density near to the Sun was approximately constant, but that it must eventually fall off with increasing distance. He also found that the rate of decrease in the number of stars per unit volume of space varied, being rapid in the direction corresponding to the galactic poles and slow in the direction of the galactic plane. The star distribution was essentially that of a strongly flattened ellipsoid of diameter 40,000 light years. Kapteyen also found that the proper motions displayed by the stars were not random, but rather showed a distinct preference for running in one of two specific directions.

It soon became clear that the proper motion data not only indicated that all stars are in motion but that the entire galactic system must be rotating. Dutch astronomer Jan Oort worked out the early theory in detail and in the mid-1920s argued that the center of the motion for the stars must be located about 18,000 light years from the Sun and that the mass interior to the Sun's orbit must be at least 60 billion solar masses. Indeed, Oort commented, there was more mass interior to the Sun's orbit than could be accounted for in terms of observed stars.

That the Sun was not located at the center of the stellar distribution presented an entirely new cosmological picture. Several studies, all coming to fruition in the early 1920s, proved this result beyond reasonable doubt. American astronomer Harlow Shapley studied the distribution and distances to globular clusters, and deduced that the galactic center must be located about 60,000 light years away in the direction of the constellation Sagittarius. Shapley also believed that the galaxy was much larger than had previously been thought; indeed, he estimated the diameter to be of order 300,000 light years.

We now know that while Shapley's theoretical analysis of the globular cluster distribution was sound and correct, they do orbit the galactic center (like a swarm of angry bees moving about a disturbed hive), but his distance estimates were far too generous. The galaxy, and the Sun's distance from its center, are much smaller than he deduced. The current, canonical number for the Sun's displacement from the galactic center is 8,000 parsecs (about 26,100 light years), and its orbital period (the galactic year) is of order 230 million years. The present-day estimate for the size of the galaxy is not far off from Shapley's estimate (about 300,000 light years across), but this is now taken to include the (assumed spherical) dark matter halo (Fig. 5.4).

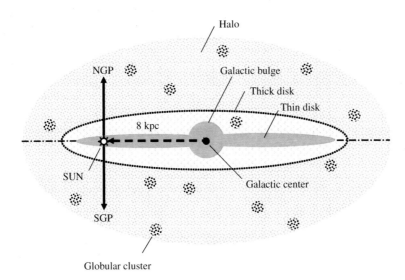

Fig. 5.4 A schematic side-on view of our Milky Way Galaxy. The thin and thick disks, the halo, the globular clusters, and the galactic bulge regions are composed of stars with distinct chemical, orbital, and age characteristics. The north and south galactic poles are indicated by the *arrow*

Unraveling the Nebula Mystery

Ever since the first telescope was used to survey the heavens, astronomers have observed strange patches of diffuse emission in the sky. Some of these light smudges were irregular in shape, others were smooth and rounded, and yet others showed tendril-like appendages. William Hershel, of course, saw these nebulae but vacillated in his opinion as to their origin. Initially he thought of the nebulae as distant accumulations of stars, "telescopic Milky Ways" he called them. Indeed, Herschel suggested that we might even be seeing star clusters in the process of forming, moving together through their mutual gravitational attraction, introducing for the very first time the idea that large complexes of stars might actually change over time – and implying in the process that they had a birth and end phase as well.

Later in life, however, Herschel discovered (and named) planetary nebulae – in which a bright central star is surrounded by a faint surrounding nebulosity (which Herschel thought reminded him of planetary disks – Fig. 5.5). Since the diffuse light in such planetary nebulae could not be made up of faint stars, unless they were exceptionally small, Herschel argued that he was either seeing a nebulosity in which the central object was not a star, or that he was seeing a star surrounded by a "shining fluid" of unknown origin and composition.

Just as astronomers in the modern era have introduced the idea of dark matter and dark energy, Herschel was not afraid of discussing the possibility that the universe might contain matter with properties that were far stranger than anything that might

Fig. 5.5 The ghostly disk of Abell 39, a planetary nebula located some 7,000 light years away in the direction of the constellation of Hercules. (Image courtesy of WIYN/NOAO/NSF)

be found on Earth. Eventually, Herschel decided that the most likely answer to the problem was that he was seeing some form of ordinary star surrounded by a strange, self-luminous fluid or gas – a partially correct summation, as it turns out, in the case of planetary nebulae, but an incorrect one for nebulae as a whole.

The year was 1845. Rising like some giant cannon above the lush green fields of Birr Castle, Ireland, was now the world's largest telescope. Sandwiched between two giant walls of massive masonry swung a gargantuan tube housing a 72-inch diameter mirror – the heart of the new telescope built for the Third Earl of Rosse. It was the leviathan of Parsonstown.

Far from an ideal location for conducting astronomical observations, the rainy and overcast Irish weather severely limited what could be achieved with the earl's new technical wonder, but this was the telescope that first revealed to human eyes the spiral structure (Fig. 5.6) of at least some of the faint nebulae that had so puzzled Herschel.

Fig. 5.6 Lord Rosse made the first detailed drawings of the spiral structure within the aptly called Whirlpool Galaxy. This Hubble Space Telescope image reveals that the nebula is actually two interacting galaxies. (Image courtesy of NASA)

Not only were many nebulae beginning to be resolved into strange shapes during the late nineteenth century, but more and more of them were being discovered. By 1888 the *New General Catalogue* compiled by John Louis Dreyer contained information on some 8,000 nebulae. As the twentieth century began to unfold, however, the mystery of the nebulae became a pressing one. What were they, and what was their relationship to the Milky Way?

Two ideas predominated the thinking at that time, and during the early 1920s the proponents for each idea went toe-to-toe in what has misguidedly become known as the Great Debate. Indeed, no actual public debate on the topic ever took place. What really occurred was a public reading of two prepared written statements of position. The first explanation, championed by American astronomer Heber Curtis, held that the spiral nebulae were external galaxies – that is, separate "island universes" (a term actually introduced earlier by philosopher Emanuel Kant in 1755) composed of billions of stars that were located at vast distances from the Milky Way. The second explanation was championed by Harlow Shapley, who argued that the nebulae were in fact part of the Milky Way, which itself was one vast "continent universe."

One of Shapley's key reasons for believing that the universe was one giant stellar system related to the fact that the vast majority of spiral nebulae were observed towards the galactic poles – that is, at right angles to the plane of the Milky Way. In sharp contrast, Shapley noted, no spiral nebulae could be seen in the galactic plane. Observationally Shapley's point was, and still is, correct, but once again it was the ISM and the dimming of starlight by interstellar dust that confused the issue. With a growing understanding of the structure and content of the ISM and how it affected starlight, it was eventually realized that Shapley's distance measures were overestimates, and his "continent universe" model fell into disfavor. Its final demise came about in 1924 when Edwin Hubble (whom we encountered in Chapter 4 and will meet again in Chapter 6) published a research paper containing an accurate measure for the distance to the Andromeda Galaxy.

Edwin Powell Hubble was not an easy man to get along with. Headstrong and opinionated he tended to dominate discussions and was less than fair in acknowledging the help he received from other researchers. Such are the ways of some human beings. With little doubt, however, Hubble revolutionized our thinking about the distribution of galaxies and the large-scale structure of the universe. Using the newly completed 100-inch Hooker telescope built at Mt. Wilson in California, Hubble began in the early 1920s to study the faint stars located within the spiral arms of several spiral nebulae (galaxies as we now call them). Hubble struck pay-dirt when he identified a number of Cepheid variables within the Andromeda nebulae, and since these stars displayed a reasonably well known relationship between maximum brightness and pulsation period, he was able to argue that the Andromeda Galaxy must be located about 275,000 parsecs away (the distance is actually much greater than Hubble's deduced number and is more like 778,000 parsecs away from the Milky Way). The distance that Hubble deduced for the Andromeda Galaxy was unimaginably vast, and it clearly indicated that the spiral nebulae must in general be distinct, individual galaxies at very remote distances. A truly large universe was beginning to open up to the gaze of astronomers.

The Galaxy Zoo

The spiral structure that was first recorded by the Third Earl of Rosse with his leviathan is just one of the many varieties of shapes displayed by distant galaxies. Some are just chaotic jumbles of stars, dust, and glowing gas; others are football (both American football and soccer ball) shaped, and yet others show looping trail-like features. Edwin Hubble, again, pioneered the study of galaxy morphology, and

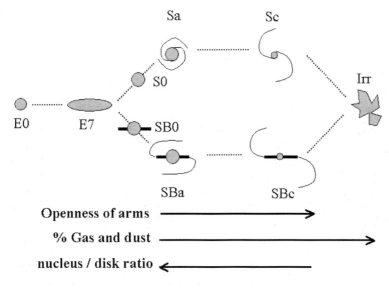

Fig. 5.7 Edwin Hubble's tuning fork diagram of galaxy classification

he derived a scheme that has been used ever since, although it is now recognized as being far from an ideal description scheme.

Hubble's tuning fork diagram for galaxy classification is illustrated in Fig. 5.7. As the diagram shows, there are broadly speaking three galaxy types: elliptical, spiral, and irregular. The elliptical galaxies can vary in appearance from being highly elliptical (almost cigar-like in shape) to spherical. The spiral galaxies are divided into two main classes, the ordinary spirals and the barred spirals. Each spiral galaxy class is then sub-divided according to the openness of its spiral arms and the central nucleus to disk diameter ratio.

Finally there is the catch-all group of irregular galaxies that contains all those galaxies that don't fit into the spiral or elliptical class. A hybrid spiral type is also indicated on the diagram in Fig. 5.7 – the lenticular (S0-type) galaxies. There is still some considerable debate as to exactly how these latter galaxies acquire their appearance; they show an essentially flattened disk (like that of the spirals), but there are either no spiral arms visible or they are so tightly wound up that they have lost their individual appearance. It appears that these particular galaxies are predominantly composed of old stars and that they are no longer undergoing any form of active star formation.

Data from some surveys indicate that about 30% of all galaxies are spiral, 10% are elliptical, and 50% are irregular. That half of the galaxies fall into the irregular class tells us something important about the environments in which galaxies evolve (as we shall see later), and while Hubble's original idea that galaxies slowly changed from being elliptical to spiral and then irregular is no longer held to be true, this is not to say that galaxies don't change their appearance over time.

Observing from within the disk of the Milky Way Galaxy it is difficult to determine exactly where our galaxy fits into Hubble's classification scheme. Radio telescope surveys of the interstellar gas clouds, as well as optical surveys relating to the location of newly formed, highly luminous massive stars, have revealed that the Milky Way Galaxy is a spiral galaxy. More recent studies at infrared wavelengths, made with NASA's Spitzer Space Telescope, further indicate, however, that the stars towards the galactic center tend to display a flattened, bar-like distribution, and accordingly it is currently suggested that the Milky Way is a SBc galaxy. The Andromeda Galaxy, in contrast, is classified as an Sb galaxy.

The Local Group

Galaxies, like stars, are gregarious by nature. It is a rare thing indeed to find a truly solitary galaxy. The Milky Way is no exception to this rule, and along with Andromeda it is accompanied in its journey through the cosmos by some 40 other companions, collectively called the Local Group. The Milky Way Galaxy and Andromeda are the largest and most massive members of the Local Group, but which is the more dominant galaxy is still not clear. Most of the Local Group members are low mass, low surface luminosity elliptical and dwarf elliptical galaxies, and there are at least a dozen or so irregular galaxies. Figure 5.8 shows a schematic map of the Local Group distribution.

Local Group galaxies, because of their relative closeness, undergo a complex gravitational dance around each other, and many mergers and disruptive encounters must have taken place during the past 10 billion years. We certainly see evidence of our Milky Way Galaxy gravitationally stripping a number of its close satellite galaxies of gas and stars. Indeed, the Magellanic Clouds, which are visible to the naked eye, and the irregular shaped Canis Major dwarf galaxy (the closest companion, in fact, to the Milky Way, at an estimated distance of just 25,000 light years) are presently in the process of being slowly ripped apart. And, indeed, the Andromeda Galaxy is presently moving towards the Milky Way at a speed of about 120 km/s. The time of closest encounter (and possible merger – the full dynamics of the encounter are not as yet fully known) is currently set for about 5–6 billion years from the present. It promises to be a grand show.

The observed galaxy morphologies and their locations within the Local Group clearly tell us that galaxy–galaxy interactions can take, and indeed have taken, place. Although the separation between our Sun and the nearest star (Proxima Centauri) is about 20 million solar radii, only 20 galaxy diameters separate the Milky Way from Andromeda. Such typical spacing indicates that while two individual stars will very rarely, if ever, collide head-on with each other, collisions between galaxies must be quite common. Indeed, the prevailing view concerning the origins of the observed galaxy types and the reason for the form of the Hubble tuning-fork sequence (Fig. 5.7) is largely determined by cluster environment and the history of past gravitational encounters.

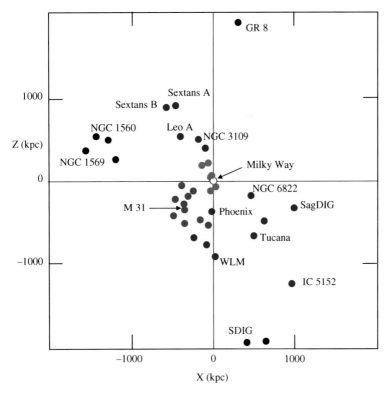

Fig. 5.8 The Local Group of galaxies extends over a region some 2 million parsecs across. In this diagram the Milky Way Galaxy is at the center and the *red dots* correspond to its satellite galaxies. The *green dots* indicate the location of the Andromeda Galaxy (M31) and its satellites

Galaxy Clusters

Compared to the many hundreds of galaxy clusters that have now been located and mapped out by astronomers, the Local Group is a flyweight. Although the Local Group contains perhaps 50 galaxies at most (allowing for a number of smaller, low surface luminosity galaxies that have yet to be found), other clusters can contain many thousands of closely packed galaxies (Fig. 5.9). Detailed survey work on the distribution of galaxy clusters indicates that our Local Group is in fact an outlier member of the much more extensive and much more massive local supercluster – the center of which appears to be located in the Virgo cluster of galaxies located some 150 million light years from us. Superclusters appear to be the largest cosmic structures that have formed within the universe, and they typically have dimensions of order 150–200 million light years.

Deep surveys of faint galaxies have revealed a complex filamentary distribution of galaxy clusters and superclusters. Figure 5.10 shows the distribution of some 100,000 galaxies as revealed by the 2dFGRS survey that was carried out at the

Fig. 5.9 The Coma cluster of galaxies is situated some 321 million light years distant from us and contains over 1,000 galaxies. (Image courtesy of NASA)

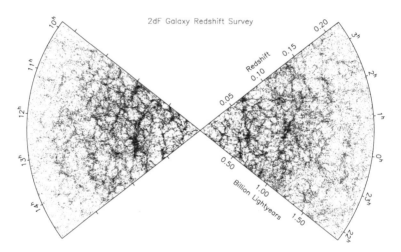

Fig. 5.10 The distribution of galaxies as recorded by the 2-degree Field Galaxy Redshift Survey (2dFGRS). This survey covers two thin wedge-shaped slices and shows galaxies out to a distance of some 2 billion light years. (Image courtesy of 2dFGRS collaboration)

Anglo-Australian Observatory. The 2dFGRS survey indicates a veritable spider's-web distribution of chains of galaxy clusters and superclusters, as well as voids – regions of apparently empty space. The deep galaxy survey observations indicate that there is no clustering beyond the supercluster scale, and this indicates that in images such as Fig. 5.10 we are essentially seeing the primordial distribution of visible matter.

We can think of this result in terms of a mixing timescale. If we take a typical galaxy within a cluster to be moving with a speed V_G of 500 km/s, then in one Hubble time ($T_H \sim 1/H_0 \sim 4.35 \times 10^{17}$ s) it might travel a total distance $D = V_G T_H \sim 2 \times 10^{20}$ km (20 million light years). The distance D is about five to six times larger than the typical dimensions of a galaxy cluster, and accordingly we might expect a galaxy to have crossed its parent cluster several times since it formed. This mixing timescale result tells us that the location of a galaxy within a cluster provides no useful information about where it originally formed.

If we extend the mixing timescale argument to superclusters, however, we then find that the galaxy clusters have not moved very far at all – in other words, the observed location of a galaxy cluster within a supercluster must be pretty much the same location in which it originally formed. The question that immediately follows this result is, how, then, did the web-like distribution of superclusters come about? The answer to this, as we shall see shortly, is intimately connected to the large-scale distribution and dynamical evolution of dark matter.

Where's the Missing Mass?

The curmudgeonly, but eminently brilliant, Fritz Zwicky (CalTech) was a remarkable lone-wolf physicist and astronomer, who was once heard to quip "I have a good idea every 2 years. Give me a topic and I will give you the idea." To perhaps prove his point, in 1933 Zwicky had a most remarkable idea; he argued in a research paper published in *Helvetica Physica Acta*, the official Journal of the Swiss Physical Society, that the dynamics of the galaxies contained within the Coma cluster (Fig. 5.9) could only be explained if there was a large proportion of unseen "dunkle (kalte) materie" ["dark (cold) matter"]. This was an incredible statement – and yet it completely failed to impress his contemporaries.

The Coma cluster is rich in galaxies, and Zwicky became interested in their individual motions. He realized that if the speeds with which the galaxies were moving could be measured (the way in which this can be done will be described shortly), then the total mass of the cluster could be derived from an application of the virial theorem. The virial theorem has many applications in astronomy and relates the total gravitational mass of a system to the velocities exhibited by its component members. All things being equal, the more matter there is in a system, the faster its components should be moving. Zwicky also knew that he could estimate the cluster mass from its total luminosity, and this should provide a check on his virial theorem calculation.

This second mass determination relies upon knowing the M/L (mass divided by luminosity) ratio for the various galaxy types (Fig. 5.7). For the Sun the M/L ratio is unity (since the mass and luminosity are expressed in solar units). For a typical spiral galaxy, however, it is found that M/L $\sim$5–10, which indicates, in fact, that most of the light we see from such galaxies must be derived from stars less massive and fainter than the Sun. Incredibly, once Zwicky had derived his two mass estimates for the Coma cluster he found that the virial mass result far exceeded that which could be accounted for by the luminosity estimate. Indeed, he found that the Coma cluster had an equivalent M/L ratio of order several hundred – at least a factor of ten larger than expected – and accordingly, he reasoned, there must be 10 times more matter in the cluster than could be accounted for in the form of visible matter.

With Zwicky's derivation of a huge M/L ratio for the Coma cluster the obvious question became, "Where is all the missing mass?" The mass, of course, was not missing. It had to be there, since its gravitational influence was driving the observed dynamics. But it was clearly locked away in some strange form that was not discernable through optical (and later radio) telescopes. Astronomers now know that most of the baryonic (that is ordinary) matter in a galaxy cluster, the Coma cluster included, is in fact contained within a hot, very diffuse intergalactic gas (revealed through X-ray telescope observations), but even including this additional mass there is still a huge amount of matter that is not accounted for.

Within 3 years of Zwicky's study of the Coma cluster being published, astronomer Sinclair Smith, working at the Mt. Wilson Observatory, revealed that the same effect was evident in the closer Virgo cluster. Interestingly, Smith speculated that the apparent difference between the dynamical mass estimate for the Virgo cluster and that expected from the mass to luminosity ratio of its constituent galaxies might be due to an intergalactic gas, but concluded his paper with, in retrospect, the far reaching words, "Whatever the correct answer, it cannot be given with certainty at this time." Smith's assessment is still true today, even though more than 70 years have passed since his paper was published.

The idea that unseen matter might exist within galaxies was not entirely new when Zwicky and Smith published their pioneering results. Jan Oort (whom we encountered earlier in this chapter), for example, had demonstrated in 1932 that our Milky Way appeared to contain more mass than could be accounted for in the form of stars. Indeed, during a speech that Oort delivered at the dedication ceremony of the 2.1-m Otto Struve telescope at McDonald Observatory in Texas in 1939, he noted that the distribution of mass in the edge-on lenticular galaxy NGC 3115 (a nebula, in fact, first observed by William Herschel) appeared to show almost no relationship with that of the light (i.e., the star distribution). The term "dark matter" appears to have been first used by Jacob Kapteyn in a 1922 research paper, published in the *Astrophysical Journal*, concerning the dynamics of stars within our Milky Way Galaxy.

Now, neither Kapteyn, Oort, nor Zwicky suspected that the missing mass that they had shown must exist was in any way different from ordinary (that is, baryonic) matter. To them it was just dark and cold, and therefore not easily observed, and to most other astronomers it was simply assumed to be an accounting problem, a problem that would eventually go away in the wake of more detailed survey work. The idea that dark matter (now properly so-called) was truly a fundamental problem did not really surface until some 30–40 years after the papers by Kapteyn, Oort, Zwicky, and Smith first appeared.

All in a Spin: Dark Matter Found

Astronomers can measure the line-of-sight, or radial speed, with which a star or galaxy is moving by means of the Doppler effect. First described in detail by Austrian mathematical physicist Christian Doppler in 1842, most of us are, in fact, familiar with the phenomena.

The most likely way to encounter the Doppler effect in everyday life is through the apparent change in pitch of an emergency response vehicle as it approaches, draws close, and then recedes into the distance again. The change in the siren's tone is an apparent change due entirely to the speed with which the vehicle is moving directly toward or directly away from the observer.

All wave-like phenomena will exhibit the Doppler effect if there is some relative, line-of-sight motion between the source and the observer, and for astronomers the phenomenon corresponds to a veritable cosmic speedometer. Indeed, the radial velocity of a star is determined by measuring the shift in the wavelength positions of absorption line features located within its spectra compared to the same absorption features seen at the telescope, where everything is at rest.

If a star is moving toward the observer, then all the absorption features in its spectra will be shifted towards shorter wavelengths, and the lines are said to be blue shifted – the apparent reduction of the wavelengths moving them towards the blue end of the color spectrum. In contrast, if a star is moving away from Earth, then all the absorption features in its spectra will be shifted to longer wavelengths, and the lines are said to be red-shifted, since they all move towards the longer wavelength (that is, red) end of the color spectrum. This Doppler change is very nicely illustrated if a star happens to be in a binary system, since under such circumstances the line-of-sight velocity will periodically shift away, and then towards, Earth as the stars move around in their respective orbits. In this case the absorption lines will be alternately red- and then blue-shifted in synchronization with the orbital period of the two stars (Fig. 5.11).

This Doppler shift technique can also be used to study the motion and rotation velocities of galaxies. In the galactic case, however, it is the group motion of all the stars at a specific distance from the galaxy center, plus the radial motion of the entire galaxy, that is measured.

The first astronomer to apply a Doppler analysis to the light from a galaxy was Vesto Slipher, who worked at the Lowell Observatory in Flagstaff, Arizona. Slipher first began studying the spectra of the Andromeda Galaxy circa 1910, and in 1913 he was able to show that it was moving towards the Milky Way Galaxy with a velocity of several 100 km/s. The following year Slipher went one step further and successfully recorded the spectra of the Sombrero Galaxy (also designated M57, and NGC 3594 – Fig. 5.12). What was particularly noticeable about the spectral lines for the Sombrero Galaxy, however, was that they were slanted (that is, not vertical). This clearly indicated that the galaxy must be rotating in our line of sight, with one half of the stars in its disk moving towards us (being blue-shifted) and the other half moving away (being red-shifted). It was an incredible finding.

Fig. 5.11 A series of
vertically stacked stellar
spectra of the binary star
system containing the white
dwarf star WD 0137-349. The
periodic shift in the dark
absorption and bright
emission lines is clearly seen
as the two stars orbit around
their common center of mass.
(Image courtesy of Pierre
Maxted, University of Keele,
UK)

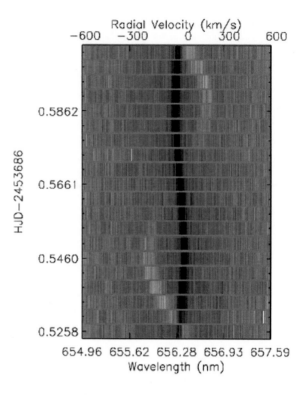

Fig. 5.12 The Sombrero Galaxy (NGC 4594). Absorption of starlight due to the interstellar dust in the disk of the galaxy is clearly visible in this Hubble Space Telescope image. (Image courtesy of NASA)

While Slipher had shown in the second decade of the twentieth century that galaxies rotate, it was not clear how the rotation velocity varied with distance from the galactic center. Radio telescope observations conducted by Dutch astronomer Hendrick Van de Hulst during the 1950s confirmed the optical analysis relating to the rotation of the Andromeda Galaxy, but the key work in establishing detailed rotation velocity profiles commenced in the 1960s. Foremost among the workers in this new field of study were Vera Rubin and W. Kent Ford, Jr. (both of the Carnegie Institution in Washington). Rubin and Ford, along with numerous co-workers, discovered over many fruitful years of observing that all of the galaxies they looked at showed a clear relationship between distinct rotation velocity versus distance from the galactic center.

The essential idea for measuring the rotation velocity variation for a galaxy is illustrated in Fig. 5.13. When observing edge-on spirals, specifically, the opening slit for a telescope's spectrograph can be set across the disk at some distance R away from its center. This will then yield a value for the rotational velocity $V(R)$ at displacement R. By adjusting the radial distance at which the spectrograph's entrance slit cuts across the galaxy, the change in the rotational velocity $V(R)$ with R can be graphed out. This is shown schematically for NGC 4565 in Fig. 5.13, where the dashed line indicates the velocity variation with distance on either side of its central nucleus. The rotation velocity shows a steady increase with increasing distance from the center and then settles down to a near constant value in the outer regions of the disk. Notice that there is a switch from motion towards (negative velocities) and motion away (positive velocities) across the galactic center. This indicates that there is indeed a true sense of circular rotation about the center of the galaxy.

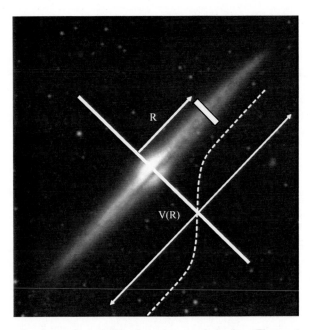

Fig. 5.13 Measuring the rotation velocity of an edge-on spiral galaxy. The spectrum is measured at various displacement values R (as indicated by the *white rectangle*) from the galactic center. The background galaxy is NGC 4565, and a schematic rotation velocity curve V(R) against R – *dashed line* – is shown in the *lower right*. (Galaxy image courtesy of ESO)

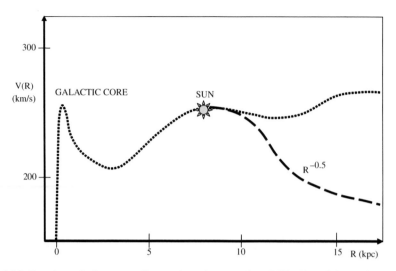

Fig. 5.14 Rotation velocity versus distance from the core of our Milky Way Galaxy. The remarkable feature of this graph is that the rotation rate does not decrease with increasing distance from the galactic center. This is a clear indication of the existence of dark matter. The *long dashed line* indicates the expected velocity variation if virtually all of the galaxy's mass were located inside the Sun's orbit

The rotation law for our own Milky Way can also be determined by studying the velocity of interstellar clouds at various distances from the galactic center. Figure 5.14 shows the velocity-distance relationship that has been derived and reveals an initial rapid increase in velocity close to the center of the galaxy and then, after a few small ups and downs, a near constant rotational velocity of about 225 km/s out to distances of at least 16 kpc.

The determination of a galaxy's rotation curve is critical to determining its mass and its mass distribution. If the velocity is known at a distance R from the center of a galaxy, then an application of Kepler's third law enables a determination of the total mass $M_T(R)$ interior to R to be made. Having once determined an estimate for M_T from the star dynamics it can then be compared to the mass of visible stars and interstellar material M^*.

The remarkable fact that astronomers have found is that in every single galaxy that has ever been studied M^* is always much smaller than M_T – that is, there is much more mass in a galaxy than can be accounted for in the terms of baryonic matter. In our Milky Way Galaxy, for example, the estimated total mass inside the Sun's orbit is of order 100 billion $M_\odot$, while the amount of mass in the form of stars and interstellar gas and dust (baryonic matter) is estimated to be of order 5 billion $M_\odot$. Remarkably, the dark matter content inside the Sun's orbit outweighs that of the stars and interstellar gas and dust by a factor of at least 20–1.

Perhaps the study that did most to raise the profile of the dark matter problem was the one published by Rubin and Ford in the June 1, 1980, issue of the *Astrophysical Journal*. This landmark paper presented the rotation curves for 21 spiral galaxies.

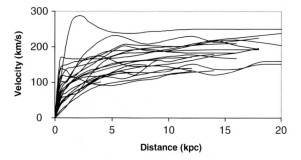

Fig. 5.15 Looking like seaweed fronds trailing in a roaring riptide, the composite set of rotation curves for the 21 galaxies studied by Rubin, Ford, and Thonnard reveals a remarkable constancy of rotation velocity with distance from the galactic center

In every single galaxy they studied it was clear and undeniable that the galaxy rotation velocities were not decreasing with increasing distance from the galactic center (Fig. 5.15). That researchers consider Rubin's and Ford's 1980 survey paper to be highly significant is revealed in the observation that, according to the NASA-operated Astronomy Abstract Service web page, it has been cited over 420 times in other research publications.

The important point to remember with respect to galaxy rotation curves is that the velocity $V(R)$ is entirely determined by the gravitational mass of the material interior to radius R. Accordingly any variations of $V(R)$ with R tell us something about the mass distribution.

Figure 5.16 illustrates two simple rotation laws. The straight line, linear law corresponds to the case where $V(R) = $ constant $\times R$, and this is the sort of law that would result if the stars were moving as if they were embedded in a rigid rotating disk. Very close to the central cores of spiral galaxies one might argue that this velocity law is evident, but it clearly doesn't apply across the entire disk of a galaxy.

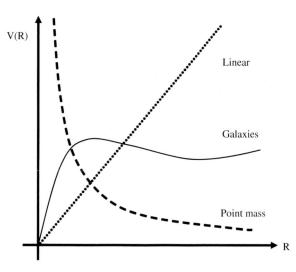

Fig. 5.16 Schematic velocity versus distance relationships. Solid disk rotation produces a linear increase in the velocity with distance; a point source mass produces a velocity that decreases as the inverse square root of the distance. Galaxy rotation curves (Fig. 5.15) are best explained, however, by the presence of a massive dark matter halo

The second, point mass law illustrates what would be seen if virtually all the mass of a galaxy were located in the center of the system. In this case the velocity is at its greatest close to the center and gradually falls away with increasing distance. Such a velocity variation law is described as being Keplerian and characteristically $V(R) = \text{constant}/R^{1/2}$. The velocities of the planets in their orbits about the Sun obey the Keplerian rule, but once again it is clear that the stars within galaxies do not. Indeed, there is no hint of the rotation velocity decreasing with distance in any of the galaxies that have so far been studied. There is no escaping the dynamical data; dark matter must exist within and around spiral galaxies, and by inference it must also exist within and surround all other types of galaxies.

The fact that the rotation curves of galaxies do not show the expected Keplerian decrease in their outer regions can mean only one thing – there must be additional mass keeping the velocity high. The simplest possible way to keep the rotation velocity constant is to have a spherical halo of dark matter distributed so that its mass increases linearly with distance R away from the galactic center. Under these circumstances $V(R)$ must remain constant.

This particular model also implies that dark matter itself must be clumped (as opposed to being uniformly distributed throughout the universe). Astronomers now believe that the dark matter halos that surround galaxies are huge and have diameters of perhaps 100,000 light years across. In addition, by studying the dynamics of satellite galaxies it is estimated that the dark matter halo surrounding our Milky Way runs to a staggering 2,000 billon solar masses. As a mind's eye approximation we might envisage the visible Milky Way as being the glowing and condensed central yoke of a massive dark matter egg.

Gravitational Lenses and Anamorphic Galaxies

The idea that the path of a light ray might be changed by a large gravitating object was first discussed by the Russian physicist Orest Khvolson in the journal *Astronomische Nachrichten* in 1924. It was a very short paper, just four paragraphs long with one simple figure, and it covered just a third of a journal page. For all of its brevity, however, this paper clearly articulated the consequences of light lensing by massive gravitational objects. The subject, however, only began to draw interest from other astronomers after Einstein himself published a paper on the topic in 1936.

Interestingly, Einstein had originally considered the effects of gravitational lensing in 1912, some 3 years before he published his epoch-making work on general relativity. The idea of light lensing is in many ways a very simple one and is a direct consequence of the requirement that under Einstein's theory of general relativity light rays must move along geodesics – that is, light travels along the shortest path determined by the geometry of spacetime.

Since large objects with large gravitational masses will warp space, the path of a light ray passing close to that object must also be altered. The light goes where

Fig. 5.17 The basic idea behind the gravitational lensing effect. In this diagram the observer O will see two distorted images of galaxy G (as indicated by the back projected paths) as a consequence of the gravitational lensing produced by the dark matter contained in the galaxy cluster C

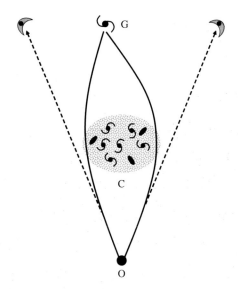

space goes. The basic idea is illustrated in Fig. 5.17. Light rays emanating from the distant galaxy G are lensed by the massive galaxy cluster C, situated between G and the observer O. According to the specific alignment between G, C, and O and the actual mass distribution within C, the lensing effect can vary from being a perfect circle to producing several distinct images of G, or to the generation of a highly distorted and stretched out image of G. Astronomers have now observed all of these consequences of the lensing effect. Einstein ring images, as they are now called, when the light from the background galaxy is transformed into a annulus of light, are relatively rare, but distorted arc-like images are very common (Fig. 5.18).

Fig. 5.18 The cluster Abell 2218 is crisscrossed with numerous gravitationally lensed and distorted images of much more distant galaxies. (Image courtesy of NASA)

Before moving on to see how astronomers make use of the gravitational lensing effect, let us consider a brief but enlightening (and hopefully) artistic aside.
Hans Holbein the Younger's painting of *The Ambassadors*, produced in 1533, has
intrigued and puzzled art historians as well as the viewing public for centuries
(Fig. 5.19). It is a wonderful painting, and it is absolutely crammed full of astronomical imagery – showing among other instruments a celestial globe, several
polyhedral sundials, a torquetium, and a quadrant. Scholars are still divided as to
the exact allegorical meaning of the painting, but what has perhaps drawn the greatest amount of interest in Holbein's masterpiece is the strange diagonal shape that
cuts across the lower part of the painting. It is, at first glance, an ugly blemish, and it
appears as if Holbein has deliberately set out to spoil his painting. And yet, there is
something compelling about the unsightly shape. It draws in our eye, and the brain
struggles to interpret its form; it is familiar and yet unfamiliar.

The unsightly addition that Holbein has added to his painting is no rude blemish, and in fact it adds further mystery to the painting's intended meaning. It is an
anamorphic image of a human skull (Fig. 5.20). From the normal viewing perspective of a painting the skull is an ugly maculation, but viewed from the right direction
it becomes a work of art, hidden, as it were, within the greater picture. The unsightly
scars and distorted arcs that crisscross the image of the galactic cluster Abell 2218
(Fig. 5.18) and many other such galactic clusters are similar to the anamorphic skull
imbedded in Holbein's *The Ambassadors*. When viewed in the correct manner the

Fig. 5.19 The Ambassadors by Hans Holbein the Younger. The painting was commissioned by
French diplomat Jean de Dinteville (shown to the *left*), and it shows his friend Georges de Selve,
Bishop of Lavaur (to the *right*). A distorted (anamorphic) image of a human skull is cast across the
foreground of the painting (see Fig. 5.20)

Fig. 5.20 By taking a picture in the direction along the diagonal of the anamorphic image in The Ambassadors painting, at a low angle to the canvas, a perfectly rendered human skull is revealed. Art historians are still undecided as to how Holbein managed to project the profile so exactly

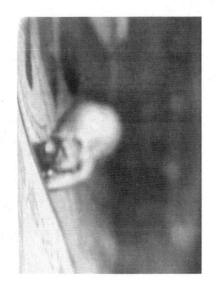

streaks should reveal normal shaped galaxies. Just as we have to adjust our viewing angle of Holbein's painting to find the image of the skull, by developing the correct anamorphic mirror to unravel the streaks produced by gravitational lensing so astronomers have been able to map out the distribution of dark matter.

The idea that we learn from Holbein's *The Ambassadors* is that anamorphic distortions can be unraveled, under the right viewing conditions, to produce a clear and recognizable image. By analogy, in the case of gravitational lensing distortions, we learn that the distorted images of distant galaxies can be unraveled to reveal the distribution of dark matter within galactic clusters. Incredibly, the diaphanous and ghost-like corpus of dark matter can be given substance and form by studying the distorted galaxy images that it produces.

COSMOS, the Cosmic Evolution Survey, is centered on a high-resolution, two-square degree sky mosaic constructed from nearly 600 overlapping images obtained with the Hubble Space Telescope. Combining the talents and expertise of nearly a hundred astronomers from a dozen countries, COSMOS is really a massive collaboration and combination of data archives obtained at visible, infrared, and X-ray wavelengths by both ground- and space-based telescopes.

Of order 2 million galaxies have been identified within the 2-degree sky column of the COSMOS field, and Richard Massey (CalTech) along with 19 co-workers have recently studied and analyzed a subset of 500,000 of them. Specifically Massey and team have studied the distortions in the gravitationally lensed galaxy images and used these to reconstruct the dark matter distribution in the line-of-sight of the COSMOS survey (Fig. 5.21).

Analysis of the COSMOS survey reveals that dark matter is both clumped and filamentary, its bulk apparently distorted by the battle between gravitational collapse and cosmic expansion. One particularly important result from the COSMOS

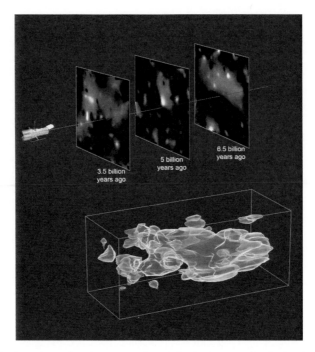

Fig. 5.21 The three-dimensional reconstruction of the dark matter distribution in the COSMOS field. (Image courtesy of NASA)

gravitational lensing survey is that dark matter is distributed in the form of a fil-amentary scaffolding, about which baryonic matter (galaxy clusters specifically) has accumulated. Indeed, comparing the locations of the dark matter concentrations with XMM-Newton Satellite X-ray images of the same field reveals a near perfect one-to-one correspondence with the positions of galactic clusters. Baryonic matter accumulations (galaxies, clusters of galaxies, and gas), in other words, represent just the luminous tips, shining beacons, if you will, of much larger and much more massive dark matter structures.

Although it is now clear that visible matter congregates in those regions where dark matter concentrations have formed, it is nonetheless possible that the two can become separated. The evidence for this has recently been found by com-bining the Hubble Space Telescope and Chandra X-ray Telescope observations of 1E 0657-56 – more conveniently known as the Bullet cluster (Fig. 5.22). This struc-ture reveals the remarkable after effects of a collision between two galaxy clusters 150 million years ago. During the collision of the two clusters the stars, galaxies, and dark matter have hardly interacted at all, but the more extensive intercluster gas has become stripped, shocked, and displaced. Intriguingly, what the Bullet cluster reveals is that the dark matter distribution, mapped out by gravitational lenses, is displaced away from the dominant baryonic, hot gas, matter component revealed by its faint X-ray glow.

Fig. 5.22 The Bullet cluster. This composite optical and X-ray image reveals the locations of dark matter (*diffuse light grey*) and hot gas (*dark grey at image center*) after two galactic clusters have collided. (Image courtesy of NASA)

Some Dark Matter Candidates

Dark matter exists. There is virtually no way of avoiding this conclusion (although see later in this chapter). It can be mapped out on the large scale by gravitational lensing, and it can be deduced from the rotation curves of galaxies and through the dynamical gyrations of galaxy clusters. It is there (well, so agree the majority of researchers), and while inherently ghost-like it seems almost tangible.

Although no universally accepted direct experimental detection of dark matter has as yet been announced (again, see later), a good number of theoretical particle candidates have been put forward to explain the dark matter phenomenon. One might even say there is an embarrassment of possible candidates. Identifying dark matter takes us well beyond the Standard Model of particle physics, and resolving what it actually is will have a profound influence on our understanding of both the microcosm and the macrocosm – within the very large is the very small, as all medieval philosophers well knew.

The Neutralino

The currently favored supersymmetry (SUSY) extension of the Standard Model predicts that all of the known particles have superpartners with large masses. Among the many SUSY possibilities for dark matter candidates is the neutralino. The current predictions are that the neutralino is stable over timescales comparable to the age of the universe, and that, remarkably, it is its own antiparticle. This latter quality of the neutralino, it has been suggested, is one way in which its presence might be directly detected – albeit with great difficulty.

The neutralino falls under the category of what are known as weakly interacting massive particles (WIMPs), meaning that it is the weak nuclear force that mitigates any interactions and accordingly such particles only very rarely notice that ordinary matter even exists. The neutrino is another WIMP that physicists have only recently been able to show has a non-zero mass and which undergoes flavor oscillations between the three neutrino types (recall Chapter 1 and see Fig. 3.1).

The neutrino is already a recognized dark matter particle, but it alone cannot account for all of the observed dark matter mass. In addition, neutrinos are generally generated in highly energetic processes, such as in the supernova explosions of massive stars (see Chapters 6 and 7), and accordingly they have near relativistic speeds. With such high speeds neutrinos cannot reasonably be expected to clump together in order to form the dark matter concentrations that the sky surveys, such as COSMOS, reveal. Effectively, therefore, the neutrinos make up a hot dark matter component. The slower moving neutralinos, in contrast, are predicted to make up the bulk of a cold dark matter component, a component that can clump due to gravitational settling and that can explain the large-scale distribution of visible matter.

There are a large number of experiments currently being developed to look for cold dark matter in the form of WIMPS (some of which will be described below). Most such experiments rely on detecting the brief flashes of light that are emitted when an atom within the detector medium is excited by a WIMP-scattering interaction. A few research groups have already claimed detection of dark matter events, but the field is alive with fierce competition, and counter claims and flat contradictions abound. As an old saying goes, extraordinary claims require extraordinary evidence. The data available to date has not convinced all researchers that dark matter interactions have been recorded.

Whatever dark matter turns out to be, it is clear that there must be a great abundance of it in the universe. Assuming a putative dark matter particle has a mass 100 times greater than that of the proton, then it has been estimated that something like 10 billion of these particles must pass through every square meter of Earth every second. A typical human being has a cross-section area of about 1 m, so during the past 2 min, the time required to read a couple of pages of this book, over a trillion dark matter particles have passed through your body. This might seem like a very large number, but in the same time interval 10,000 trillion solar neutrinos have also passed through your body. That's a whole lot of WIMPS all completely ignoring you – and, of course, mostly ignoring any dark matter detection experiment that happens to be running as well. Pinning dark matter down in the laboratory is going to require both cunning and guile.

In addition to looking for laboratory examples of WIMP interactions with ordinary matter, another possibility is to investigate the self-destructive properties predicted for some dark matter candidates. The neutralino, for example, is its own antiparticle, and accordingly if two neutralinos come into contact, they will be annihilated, producing a shower of ordinary particles such as electrons, protons, and positrons. Such decays, it has been argued, might leave a measurable signature within the cosmic ray background (as will be described shortly).

Perhaps the main rival to the neutralino as the leading dark matter candidate is the axion. This particle has been invoked to explain the so-called strong CP violation, and most theorists believe that this particle must exist, the only problem being that its mass is currently unknown. Importantly, however, it is predicted that in the presence of a very strong magnetic field an axion should convert into a photon, and this leads to the possibility of experimental detection.

Looking for MACHOs

It would be far too naive to think that all of the dark matter observed in the galaxy and the greater universe is in the form of one single exotic particle. Within our Milky Way Galaxy, and all other galaxies for that matter, there is a baryonic dark matter component. Indeed, candidate Massive Compact Halo Objects (MACHOs) have been identified in gravitational lensing surveys of the halo of our galaxy. The detected objects are most likely old, cold, and very low luminosity white dwarfs. (These objects will be discussed in detail in Chapter 6.) Other MACHO candidates include cold Jupiter-mass planets, black holes, neutron stars, and low luminosity brown dwarfs (substellar objects with masses of between about 1/10 that of the Sun and about 15 times that of Jupiter).

We know that these objects do exist within our galaxy, and there is no reason to believe they don't exist in others. The MACHO mass contribution to the dark matter compliment, however, is likely to be relatively small, but some studies suggest that as much as 20% of the dark matter mass within our galaxy falls within the MACHO domain.

DAMA Finds WIMPS, Maybe

Situated deep underground in the Gran Sasso National Laboratory in central Italy are a suite of DArk MAtter (DAMA) experiments, utilizing various detector materials. The key aim of each experiment has been to detect the rare flashes of light produced by the recoil scattering of detector nuclei after interacting with a dark matter WIMP.

The main focus of the DAMA project at the present time centers on the Large sodium Iodide Bulk for RAre processes (LIBRA) experiment (Fig. 5.23). This experiment consists of 25 blocks of sodium iodide, stored in a highly pure nitrogen atmosphere and encased in multiple layers of copper and lead shielding – not to mention the 1,400 m of dolomite rock above the detector hall. The sodium iodide blocks are monitored by a series of photomultiplier (PM) tubes, sensitive electronic

Fig. 5.23 A technician inspects the copper and lead shielding that surrounds the sodium iodide detectors at the heart of the DAMA/LIBRA experiment. (Image courtesy of the LNGS)

eyes that continuously watch and wait to record the brief flashes of light that betray potential dark matter interactions.

The search for dark matter interactions, however, is not a game for the hasty of heart to play, and of the 1,000 trillion dark matter particles that will pass through the DAMA/LIBRA detector per day perhaps one interaction every few days will actually be recorded by the PM detectors. In the same time interval many background, that is non-dark matter related, events will have also been recorded. The problem with all such monitoring experiments is separating the signal from the noise.

The multinational, multi-university DAMA research group published its first results in the late 1990s and claimed that an annual modulation in the dark matter detection rate had been recorded (Fig. 5.24). This annual modulation is an intriguing, indeed, incredible result, but one that is not universally accepted. The modulation, the DAMA group argues, is due to a velocity variation related to Earth's motion around the Sun. Specifically, the argument goes, the characteristic interaction velocity between the detector and any dark matter particle will be modulated by ± 30 km/s, Earth's orbital velocity, on a 6-month timescale, and this will produce a measurable variation in the interaction energy.

During the course of a year the dark matter interactions should have slightly higher energies in the summer months, when Earth's orbital velocity adds to that of the Sun's motion around the galactic center, and the energies should be slightly smaller in the winter months, when Earth's orbital velocity is opposite to that of the Sun's galactic motion. The DAMA/LIBRA experiment cannot identify what kind of dark matter particle has been detected (other than it being a WIMP), but the collaboration continues to argue that a clear modulation of the dark matter signal has been recorded over a 7-year time span. Other experiments are trying to recreate

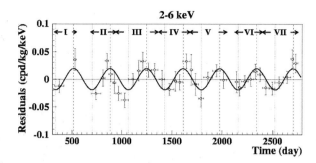

Fig. 5.24 Modulation of the dark matter signature detected by the DAMA researchers. The data points correspond to the number of events recorded per unit time, per unit detector mass, and specified energy range (in this case, 2–6 keV). The data points, when plotted against the dates over which the experiments were run, show a distinct wavelike variation. The expected yearly modulation due to Earth's motion about the Sun is shown by the *solid sinusoidal line*. (Image courtesy of DAMA)

the DAMA results, and the search for dark matter, with other kinds of detectors, continues apace.

CDMS Sees Two, Well, Maybe

The Cryogenic Dark Matter Search (CDMS) experiment sits 800 m below ground in the Soudan mine in northern Minnesota. The experiment is essentially a super-sensitive heat detector designed to record the very specific signatures that indicate a WIMP has interacted with one of its 30 super-cooled germanium-silicon detectors. The experiment has been running since 2003, but data results announced on December 18, 2009, caused a definite buzz of excitement within the physics community. Reporting on observations gathered between July 2007 and September 2008, the CDMS researchers argued that two of their recorded events had all the hallmarks of being due to WIMP interactions. As with all such experiments, however, the CDMS results are based upon a statistical analysis of the data, and the possibility that the two events are due to background noise (radioactive decay or cosmic ray induced events) cannot be ruled out. Yet it is still far from clear if the first definitive dark matter detections have been made. The search continues.

Bubbles at COUPP

How do you catch sight of something that can't be seen? This is the problem posed by dark matter, and it's the sort of conundrum that Charles Dodgson would have probably made great play of. Indeed, just like the smile of Dogson's Cheshire Cat in *Alice's Adventures in Wonderland*, one can only study the "body" of dark matter indirectly. Experimental physicists literally must design their equipment to look for the remnant "smile" of dark matter, left behind, as it were, in the visible world on those rare occasions when an interaction with ordinary matter actually takes place.

Although it is probably unreasonable to expect dark matter to produce something as complex as a smile, it is not unreasonable to hope that dark matter might be made to blow a bubble – a prospect, no doubt, that would have also appealed to Dodgson. Rather than being some doggerel in a fantasy novel, however, this is exactly the effect that scientists from Fermilab, the University of Chicago, and Indiana University are using to test for the existence of certain kinds of dark matter WIMPS.

The Chicagoland Observatory for Underground Particle Physics (COUPP) was established in order to allow physicists to develop and perform particle physics experiments in a belowground environment that provides shielding from the natural background interference of cosmic rays. Such shielding, however, will have virtually no effect on WIMPS, and accordingly, 350 ft underground at Fermilab is where the dark matter experiment resides – unhindered by spurious cosmic ray interactions. The key component of the experiment is a dusted down, but fully modernized, piece of old technology – the bubble chamber (Fig. 5.25).

The working principle behind a bubble chamber is to place a detector liquid into a superheated state. This is achieved by first adjusting the surrounding pressure so that the detector liquid is just below its boiling point. Once in this condition a very sudden drop in the pressure leaves the liquid in a highly unstable superheated state. It remains liquid even though it is above its theoretical boiling point. Once a liquid has been superheated, however, any small disturbance will trigger the boiling process, and in the case of atomic-scale interactions this will trigger the formation of small bubbles that will rapidly grow in size and can be photographed.

The first version of the COUPP dark matter detector employed 1.5 kg of superheated iodotrifluromethane (a fire-extinguishing fluid with the chemical formula CF_3I) as the detecting fluid. The operating principle behind the experiment is that if any dark matter particle interacts with one of the atoms in the superheated CF_3I then the resultant recoil motion should trigger the formation of a bubble. Only a single bubble, rather than a train of bubbles, is expected in a dark matter interaction, since the probability of interaction is so very small (Fig. 5.26).

The first round of experimental data was published by the COUPP group in February of 2008, and its results were at odds with those derived by the DAMA/LIBRA researchers. Indeed, the debate between the two groups has proceeded with some degree of hostility – a clear indication of both the importance and competitive nature of the dark matter search. According to the initial COUPP study, which gathered data for about a year, if the DAMA results were indeed true dark matter detections then the superheated CF_3I detector should have recorded hundreds of single-bubble events. It turns out that they observed none. No doubt eventually the differences of opinion and interpretation will be resolved, one way or the other, but one thing is for certain: more experimental work is required. Indeed, plans for a much larger dark matter detecting bubble chamber containing 30 liters of CF_3I are already well advanced at Fermilab.

Fig. 5.25 At the heart of the COUPP detector is a small bubble chamber, shown here removed from and situated above its outer casing. (Image courtesy of Fermi Lab)

CHAMPs and SIMPs

Is it possible that dark matter carries an electrical charge? Most theorists believe no, but Leonid Chuzhoy and Rocky Kolb, both at the University of Chicago, have recently resurrected this previously rejected idea and suggest that there might be a CHArged Massive Particle (CHAMP) component to the dark matter particle pantheon.

If CHAMPS really do exist, then there are a number of interesting consequences that might follow. If negatively charged, for example, they might bond with ordinary atoms and produce supersized molecules that could be detected by their excessive weight. Indeed, the predictions suggest that CHAMPs might have masses as high as 100,000 times that of the proton.

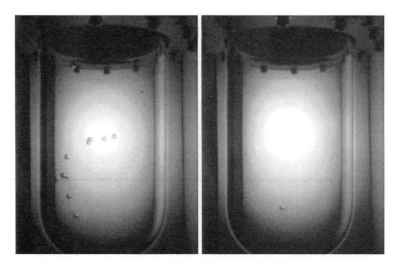

Fig. 5.26 Bubbles form and expand in the superheated CF_3I fluid of the COUPP detector. It is expected that dark matter interactions will result in the formation of a single bubble. (Image courtesy of Fermilab)

Even if CHAMP-baryonic molecules can't form, it is still possible that they might collide and interact, producing in the process a distinctive electromagnetic signal observable to terrestrial telescopes. The theoretical details have yet to be fully worked through, but one potential problem concerning the detection of CHAMPS has already been raised. Since they are charged their distribution will be controlled not just by gravity but by the shape and extent of a galaxy's magnetic field. This interaction could result in CHAMPs being excluded from the disk of our galaxy, and accordingly from any satellite or Earth-based experiment. In addition, the very high mass predicted for CHAMPs also means that they are far too heavy to be produced in the LHC for direct study. It is far from clear at the present time if CHAMPs really do exist, and even if they do, a pre-print publication posted in December of 2008 by Javier Sanchez-Salcedo and Elizabeth Martinez-Gomez, of the Universidad Nacional Autónoma de México, has argued that at best they might contribute up to 1% of the total dark matter mass of our galaxy's halo.

Strongly Interacting Massive Particles (SIMPs) are conjectured to be electrically neutral particles that can interact with ordinary matter. Although the details of their properties are unclear, it is argued that they can bind, via the strong interaction, to the nuclei of ordinary matter. In this manner their presence might be revealed experimentally through the detection of anomalously massive isotopes of otherwise ordinary atoms. A search for SIMP anomalies in fragments from the Canyon Diablo meteorite, samples of gold foil flown in Earth orbit for nearly 6 years as part of NASA's Long Duration Exposure Facility (LDEF), and a gold sample from the beam dump at Brookhaven's RHIC experiment was carried out by Daniel Javorsek II (then at Purdue University) and collaborators in early 2000–2001. The experiments, however, revealed no SIMP candidates.

PAMELA Finds an Excess

The Payload for Matter-Antimatter Exploration and Light-nuclei Astrophysics (PAMELA) spacecraft was blasted into Earth orbit in June of 2006 (Fig. 5.27). Carried onboard the satellite is a sophisticated cosmic ray detector that has been designed to carefully distinguish between the number of electrons and positron cosmic rays encountered. The rationale for the mission is based upon the theoretical prediction that dark matter particles, such as the neutralino, are their own antiparticles, and accordingly they will be annihilated during encounters. Such dark matter annihilation events will take place within the extended halo of our galaxy (recall Fig. 5.4), and, it is argued, they should therefore produce a measurable excess of cosmic ray positrons and electrons. It is the predicted excess of high-energy positrons over electrons within the cosmic ray spectrum that the PAMELA instrument is looking for.

Within a year of PAMELA's launch an excited buzz began to circulate within the dark matter research community. A positron excess had been found at exactly the predicted energy range. Tantalizing glimpses of the data were presented at various research meetings, but finally, in October of 2008, the PAMELA consortium uploaded a pre-print to the arXiv server (a database from which technical research papers submitted for publication can be viewed). In this pre-print article, lead scientist Piergiorgio Picozza (University of Roma Tor Vergata, Italy) along with

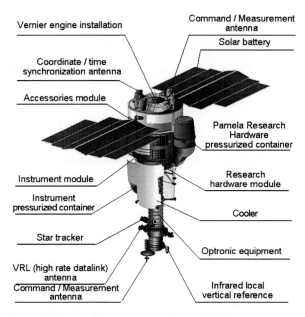

Fig. 5.27 The Resurs DK-1 satellite aboard which the PAMELA experiment is carried. The PAMELA researchers have reported an excess of cosmic ray positrons, with the excess being attributed to dark matter annihilation events in the galactic halo. (Image courtesy of the PAMELA collaboration)

50 co-authors announced that PAMELA had indeed found a significant and indisputable positron excess. This excess, they cautiously conclude, "may constitute the first indirect evidence of dark-matter particle annihilations." The caution, of course, is justified, since there are a number of other possible explanations for the observed excess besides dark matter, and these alternatives have yet to be fully ruled out. The peer-reviewed and refereed version of the research paper written by the PAMELA consortium appeared in the April 2, 2009, issue of the journal *Nature*.

Fermi's Needle in a Haystack

The Fermi Gamma-Ray Space Telescope (Fig. 5.28) was successfully launched in June 2008. It is the most sensitive gamma-ray telescope ever to be placed in low Earth orbit, and its job is to survey the entire sky in the very short, gamma-ray wavelength region of the electromagnetic spectrum.

The Fermi survey will attempt to resolve and map out the gamma-ray sky into both its diffuse and discrete components, as well as survey the sky for gamma-ray

Fig. 5.28 An artist's impression of the Fermi Gamma-Ray Space Telescope. In addition to surveying the sky at gamma-ray wavelengths, the detector aboard the spacecraft will look for signals relating to dark matter annihilation events. (Image courtesy of NASA)

bursts and transient outbursts. In addition it will also look for signatures of dark matter annihilation. These latter observations will complement those being conducted by PAMELA, but will concentrate on detecting the result of so-called suppressed annihilations in which two dark matter particles interact and decay directly into two high-energy gamma rays. Such decays will produce a distinctive line spectrum.

The most likely location where such decays will be observed is that of the galactic center, since it is there that the dark matter density is expected to be at its highest. The problem for the researchers, however, will be to distinguish the dark matter annihilation signal from the more diffuse, natural gamma-ray background. This will be a formidable task and one akin to looking for the proverbial needle in a haystack. Early results from the Fermi spacecraft detector also indicates an electron and positron cosmic ray excess, but the researchers interpret their data as being due to a local pulsar (see Chapter 7 for a discussion of these strange objects) emission rather than dark matter annihilations advocated by the PAMELA group. The debate continues.

ADMX

Phase 1 of the Axion Dark Matter eXperimant (ADMX) is located at the Lawrence Livermore National Laboratory in California, and the stated goal of the experiment is to find "the elusive axion that is a candidate for dark matter." The axion is a hypothetical particle that forms part of the detailed theory of strong nuclear interactions, although its name is derived from much more humble origins and is taken from a now-defunct brand of laundry detergent.

The ADMX has been designed around the prediction that axions, provided they actually exist, should occasionally decay in the presence of a strong magnetic field into two photons. Since, however, axions are predicted to have very low masses, the two photons will also be of low energy and therefore difficult to detect. The experiment in its Phase 1 form consists of a so-called resonance chamber that is placed within a strong magnetic field (Fig. 5.29). Highly sensitive and very low noise amplifiers then monitor the inner core chamber for the telltale microwave photons that might emerge from any axion decays.

To date no indications of any axion decays have been found in the ADMX, but the program is currently in the process of constructing a Phase 2 experiment that will be housed at Washington University. The present hope is that definitive results from the Phase 2 experiment will become available by 2011.

Euclid's Dark Map

Euclid of Alexandria lived circa 300 B.C. and has often been called the father of geometry. His great mathematical thesis *The Elements* is still famous among mathematicians to this very day.

Fig. 5.29 A research technician oversees the extraction of the cooled resonance cavity from the magnetic bore of the ADMX. (Image courtesy of the ADMX consortium)

Adopting Euclid as their muse and inspiration the European Space Agency is currently evaluating a space-based mission with the aim of mapping out the geometry of the dark universe. The currently envisioned Euclid spacecraft will support a 1.2-m telescope operating in the visible and near infrared part of the electromagnetic spectrum. The launch is provisionally scheduled for 2017, and during a 4–5-year mission the spacecraft will systematically perform an all-sky survey of the heavens.

The science objectives of the Euclid mission are straightforward and aim to probe and quantify both the expansion history and the density evolution of the universe. To achieve these goals the spacecraft's detector system will acquire thousands of images and spectra of distant galaxies as well as clusters of galaxies that have been distorted by foreground dark matter. From these images the researchers will then be able to unravel and correlate the observed distortions (recall Fig. 5.19) with cosmic distance, and thereby map out the dark matter distribution on an unprecedented scale. Full funding for this mission has not as yet been approved by ESA.

The MOND Alternative

The deduction that dark matter must exist within the observable universe is predicated on the assumption that general relativity, as originally formulated by Einstein in 1915, is a correct representation of how gravity works. Most researchers believe that this assumption is valid and correct, and indeed, general relativity has never failed to make a correct prediction under those circumstances in which it is expected to apply and where it can be tested.

What, however, are the observable consequences if gravity behaves differently than the predictions of our cherished model? We know, for example, that the formalism of general relativity must break down on the very small quantum scale, although there is no current consensus on how it should be reformulated. Likewise, it is also well known that far away from large gravitational masses general relativity reduces down to the simpler Newtonian gravitational approximation.

It is within this latter realm that some researchers have speculated that new physical phenomena, beyond previous predictions, might reside, and indeed, it has been argued that dark matter might not exist at all. A good number of alternate theories to general relativity have been published over the years, but in recent times it has been the MOdified Newtonian Dynamics (MOND) model that has received the most attention.

MOND was first proposed by physicist Mordechai Milgrom (Weizmann Institute, Israel) over 25 years ago, and it has shown remarkable resilience against the attacks by its (apparently) numerous distracters. The incredible simplicity of the argument behind MOND is breathtaking, and it applies to a realm of gravitational interactions where we have no direct experimental verification of what actually goes on. The modification that Milgrom and co-workers propose applies under those conditions in which the gravitational acceleration is very, very small (such as in the far outer reaches of a galaxy), and specifically a modification of Newton's second law of motion is invoked.

Under high acceleration conditions Newton's second law has been experimentally verified and provides a relationship between the acceleration a produced by a force F acting upon a mass m such that $F = ma$. What Milgrom proposes is that once the acceleration drops below some threshold value a_0 then Newton's second law of motion is modified to become $F = m(a^2/a_0)$. The threshold constant is not specifically defined by the theory, but it is taken to be a new fundamental constant of nature that can in principle be derived from the observations.

The important point about the MOND reformulation of Newton's second law is that it requires the velocity of all stars traveling along circular orbits at great distances away from the center of a galaxy to move with a constant velocity. In other words, it forces a galaxy rotation curve to become flat with increasing distance from the galactic center – just as shown by the observations (Fig. 5.15). Indeed, the observations of galaxy rotation curves are remarkably well described by MOND, and it appears that the threshold acceleration a below which the MOND equation "switches on" is $a < a_0 = 1.2 \times 10^{-10}$ m/s^2.

The MOND model is still considered controversial, to say the least, and critics point out that there is no specific rationale behind the requirement that Newton's second law of motion should change in the low acceleration domain. Importantly, however, the theory does makes a set of clear predictions that apply under all very low acceleration conditions, and this may pave the way for experimental testing.

The problem presented, however, is exactly how to test MOND within the controlled conditions of a laboratory or a spacecraft. Indeed, the first problem is that one will first have to move much further away from the Sun than the 1 astronomical unit (AU) orbital radius of Earth. To move into a domain where the Sun's gravitational acceleration is of order a_0 requires a displacement of at least 7,000 AU – putting the test domain well into the cometary Oort Cloud region of the Solar System. While, in principle, space-based experiments might be conducted at such distances eventually, the potential timeframe for their initiation is set far into the future.

In the meantime, one important extragalactic testbed that astronomers have identified for MOND is that of the Bullet cluster (Fig. 5.22). At issue here is the fact that the baryonic matter producing the X-ray emission has become separated from the main mass component (identified earlier as dark matter) responsible for producing the gravitational lensing. Closer to home, the author has recently looked more carefully at the possible gravitational interaction between α-Centauri A/B and Proxima Centauri, and found that Proxima's orbit could well be an important MOND test case. Proxima is located about 15,000 AU from α-Centaurus A/B, and the gravitational acceleration experienced between the two systems currently amounts to about $a_0/2$. Accordingly Proxima's orbit should be controlled by a MOND interaction.

At the present time the exact characteristics of Proxima's orbit relative to α-Centauri A/B are not fully known, but they should all be measurable to a high order of accuracy in the near future, and this will be a good test of the MOND model. Interestingly, however, the presently observed velocity of Proxima relative to α-Centaurus A/B is almost exactly that predicted by the MOND equations.

There is a great deal of intellectual prestige at stake with respect to MOND. If it turns out to be a true, then the models describing the dynamical workings of the universe will all need reformulating.

Dark Stars and Y(4140)

To date the existence of dark matter has been predominantly inferred from the study of large structures – galaxies and clusters of galaxies. With the possible exception of the CHAMPs model, however, the inferences drawn from the large-scale observations are that dark matter should be present on much smaller scales – the scale of the stars, planets, and, more importantly, terrestrial laboratories. On the stellar scale it has recently been suggested that the very first stars that might have formed in the universe, some 1 billion years after the Big Bang, were powered by dark matter annihilation.

Physicist Katherine Freese (University of Michigan) and co-workers have recently published a whole series of articles outlining the possible properties of the first "dark stars." These stars, it turns out, are likely to be stellar behemoths, weighing in at perhaps 800–1,000 times the mass of our Sun. Rather than being stabilized against collapse by fusion reactions (as our Sun is) dark stars are supported by the energy liberated by dark matter annihilations. In principle, Freese and her collaborators suggest, the primordial population of dark stars might be detectable through gravitational lensing, and it is entirely possible that some dark stars may even exist to this very day, depending upon the dark matter accretion circumstances. If such dark stars are identified, then they would provide a very useful testbed for dark matter theories, since their structure is directly related to the mass and interaction cross-section of the constituent WIMPs. At the present time, however, there are no observations of possible dark star candidates – but the search continues.

Closer to home it is entirely possible that dark matter has become gravitationally bound to Earth and the other planets within the Solar System. Indeed, Stephen Adler, working at Princeton's Institute for Advanced Study, has recently suggested that some of the anomalous heating effects observed in the Jovian planets might be due to dark matter accretion. Adler also notes that if the Sun has an associated dark matter halo then the rotation of Earth should produce a 24-h modulation in dark matter detection rates. This effect would complement and be in addition to the yearly modulation deduced, for example, by the DAMA/LIBRA experiment.

While the design and construction of the LHC has been meticulously planned, its overall success as an experimental device will most likely follow from serendipitous and unexpected discoveries. It is, indeed, the unlooked for discovery that makes experimental science so compelling. A wonderful example of this unexpected discovery was recently announced, in November 2008, by researchers working at Fermilab's Tevatron particle accelerator (Fig. 5.30) in Batavia, Illinois. The first observational hints of a new particle, dubbed rather unceremoniously Y(4140), having been discovered were obtained with the Collider Detector at Fermilab (CDF), which monitors the spray of particles produced during head-on collisions between protons and anti-protons. The new particle has a mass of 4,140 MeV (hence its ungainly name), and exists for a relatively long 20 ps before it ultimately decays into muons; this suggests some highly interesting possibilities. In particular physicists Neal Weiner (New York University), Nima Arkani-Hamed (Institute for Advanced Study at Princeton), and colleagues were keenly interested in the CDF results since the estimated mass of the Y(4140) is a near exact match for a force-carrying particle that they predict should exist as a consequence of dark matter interactions.

Time will tell, of course, if the new CDF results are real or some presently unidentified experimental artifact. But if they are true, then maybe, for the very first time, signs of dark matter being generated and interacting within a terrestrial experiment have been recorded.

The design and implantation of new experiments to detect dark matter interactions and annihilation events continues apace. It is a hot topic, with a bright future.

Fig. 5.30 The CDF with which researchers found the new and completely unexpected Y(4140) particle. (Image courtesy of Fermilab)

The CERN researchers will be playing their part in the frenzied hunt to identify dark matter candidates, and it is fair to say that great scientific fame awaits those fortunate scientists to make the first bona fide discoveries. Whether WIMPs, SIMPS, MACHOs, or CHAMPs, at the present time dark matter is profoundly mysterious to us, but incredibly it is by no means the strangest phenomenon to apparently exist in the universe. By far the most bizarre component of the cosmos is dark energy, as Chapter 6 will endeavor to explain.

Chapter 6
Dark Energy and an Accelerating Universe

Dark energy is an enigma. It could be one of the greatest discoveries of modern cosmology, or it might not exist at all – being but a cruel artifact of the presently available data. For once the theoretical physicists are stumped and collectively they can offer no explanation for what dark energy, if it is real, could be due to. The situation is challenging and remarkable, and while astronomers can point to data that suggests the possible existence of dark energy, no one can currently explain why such a strange situation should come about.

The CERN researchers are fully aware of the problem, but there is no LHC experiment that will be looking for, or indeed, is expected to reveal anything specific about what dark energy might be. In fact, there are currently no laboratory experiments being performed anywhere in the world that might provide any direct insights relating to dark energy. It is possible that the LHC might reveal unexpected new physics that could shed some light on the problem, but it will be serendipity rather than design that will pave the way.

This chapter, therefore, will have very little to say about what the LHC may, or may not reveal about dark energy, but since it is one of the hottest topics of modern physics and cosmology it has a natural place within our story. Not only this, but at the heart of the debate concerning dark energy is the observation of supernovae, the bedazzling fireballs of light produced when stars literally blow themselves apart, and it is these same supernovae that produce the cosmic rays, neutron stars, and black holes that will be the topic of the next chapter. So, we shall begin by asking what are supernovae, and then move on to consider what it is about the observational data they present that suggests the universe might just be bursting full of some type of mysterious dark energy.

The near simultaneous announcement, in the late 1990s, by two international research consortiums that the entire universe must be permeated by an all-pervading dark energy was a complete and utter surprise to cosmologists. It was a literal lightning bolt out of the blue. Remarkably, according to present-day estimates, in terms of mass-energy equivalency, dark energy accounts for 73% of the mass of the universe. Dark matter accounts for a further 23% of the total, while visible stars along with interstellar and intergalactic gas (that is, baryonic matter) complete the final 4% (Fig. 6.1). Incredibly, the stars and galaxies that we see in such abundance are the

M. Beech, *The Large Hadron Collider*, DOI 10.1007/978-1-4419-5668-2_6,
© Springer Science+Business Media, LLC 2010

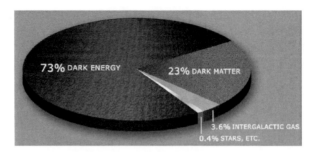

Fig. 6.1 The estimated distribution of dark energy, dark matter, stars, and intergalactic gas in the present epoch. As time progresses into the deep future the dark energy mass contribution within the observable universe will increase at the expense of both the dark matter and the baryonic matter contributions. (Image courtesy of NASA)

proverbial tip of the iceberg, and the vast majority of the mass in the universe is composed of something entirely alien to our everyday experience and understanding.

Although the appearance of observational data supporting the possible existence of dark energy came as a surprise to astronomers, this is not to say that theoreticians hadn't already anticipated its possible existence. Indeed, when the ever-inventive Einstein first turned his attention to cosmological issues in the 1920s he opened the door to the possibility of the universe being pervaded by either accelerating or retarding energies – he did this by adding an additional constant term to his equations of general relativity.

The reasons for Einstein's addition were quite straightforward, since he wanted to emulate a static universe – literally, a universe in which there was no net expansion or contraction, a universe with no beginning and no end. Einstein's thinking was entirely in line with the available astronomical data at that time, and since his general relativity equations indicated that the universe had to be either contracting or expanding (it didn't specify which), he made use of a convenient, but entirely justified, mathematical dodge and added a non-zero constant term. By introducing this constant term, denoted by the Greek symbol Λ, Einstein could fine-tune his model results to produce the requisite static cosmos.

When first proposed by Einstein in 1917 the cosmological constant made sense; the universe appeared to be static, and something had to keep it from either expanding or contracting. The days of Einstein's cosmological constant were numbered, however, and within a dozen years of his initial research paper being published American astronomer Edwin Hubble announced, in 1929, the discovery of universal expansion (to be described shortly). Realizing that he had failed to predict the possibility of an expanding universe Einstein described the cosmological constant as his "greatest blunder." If we skip forward some 65 years, however, we might now say that Einstein's cosmological constant was his "greatest wonder," since the evidence is now beginning to accumulate that Λ may not, in fact, be zero.

There is presently no generally agreed upon theory to explain dark energy. It has no obvious explanation in terms of quantum mechanics and/or standard physics. It is a genuine twenty-first century mystery. Provided the current set of observations stand the test of time, and they may not, then dark energy will have profound consequences for physics. It will also have profound consequences with respect to the appearance of the universe in the deep future.

Before addressing the more theoretical issues relating to dark energy, let us first review the chain of reasoning that has lead astronomers to suggest that dark energy is not only driving universal expansion but that it is driving an accelerating universal expansion. Our story begins close to home and deals with the question of how astronomers measure the distances to the nearest stars.

The Measure of the Stars

Gauging the scale of interstellar and then intergalactic space is far from trivial. There are many interpretational pitfalls that can bias what we see and accordingly deduce, resulting in either an over- or an underestimate of distances. Fortunately, for the nearest stars to us, distances can be determined by the direct measurement of stellar parallax (Fig. 6.2).

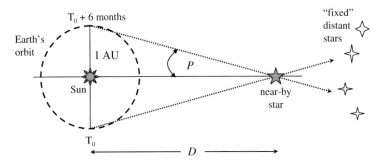

Fig. 6.2 Stellar parallax due to Earth's motion around the Sun. The angle P is the angle of parallax, and the distance to the *star* is given in terms of Earth's orbital radius, defined as 1 astronomical unit (AU) ≈ 149.6 million kilometer. By simple trigonometry the tangent of angle P is equal to 1 AU/D

The geometrical basis behind parallax measurements were well known to ancient Greek philosophers, but it required several thousand years of technological advancements before the apparent 6-monthly, back and forth shift in the relative sky location of nearby stars could be measured with any certainty. Indeed, after several premature announcements and false starts by other astronomers, stellar parallax was finally measured, between 1838 and 1840, by Carl Friedrich Gauss, Thomas Henderson, and Friedrich Struve. These three pioneering astronomers measured the parallaxes for 61 Cygni, α-Centauri A/B, and Vega, finding their half-angle shifts over 6 months to be 0.287, 0.742, and 0.129 seconds of arc, respectively [1 seconds of arc corresponds to 1/3,600 of a degree]. With these values the distances to 61 Cygni, α-Centauri A/B, and Vega are 3.48, 1.35, and 7.75 parsecs, respectively. The distance unit of the parsec is defined as being the distance at which the angle of parallax amounts to 1 second of arc, and the name parsec is based upon a contraction of the expression "*par*allax of one *sec*ond."

The closest star to the Sun is the low luminosity, low mass, M-dwarf star Proxima Centauri. This stellar lightweight was discovered in 1917 by Robert Thorburn Innis while working at the Transvaal (later the Union) Observatory in Johannesburg, South Africa. Having a parallax of 0.772 seconds of arc, Proxima is located some 1.295 parsecs away. It is curious to think that the closest star to us (after the Sun, of course) is too faint to be detected by all but the largest of telescopes; it is nine magnitudes fainter than the human eye can see, and it corresponds to the brightness that the Sun would have if viewed from a distance of 1,337 parsecs. Proxima is, indeed, one faint star.

Once the distance to a star is known, then it is transformed into a new kind of object – a standard candle. This transformation comes about since the known distance can be combined with observations of the star's energy flux, allowing its true energy output per unit time (the luminosity) to be deduced. The trick thereafter is to associate a distant independent characteristic of the calibration star (for example, its spectrum) against its deduced luminosity. In this manner any other star with the same spectra as our calibration star will have a known luminosity – that is, the luminosity of the calibration star.

Having established one set of standard candles, astronomers can then determine the distances to other objects by a bootstrap method: calibrate those stars that are

Fig. 6.3 Of the 150 or so known globular clusters surrounding our Milky Way Galaxy, NGC2808 (shown here) is one of the most massive. The cluster contains about a million stars, is located about 9,000 parsecs away, and is composed of old, low mass stars that formed about 12.5 billion years ago. (Image courtesy of NASA/HST)

close enough to have measured parallaxes against any distant independent stellar feature, e.g., its spectrum. Then use the calibration feature to recognize and determine the distance to those stars that are too far away to have their parallaxes measured. When distances become too great to measure the individual properties of stars, then calibrate larger, brighter structures such as globular clusters (Fig.6.3) and then use these objects to determine the distances to more remote globular clusters associated with other galaxies, and then calibrate entire galaxies, and so on.

By far the brightest standard candles known to astronomers are supernovae (discussed more fully later), and these can be used to measure the distances to galaxies that are situated many billions of parsecs away. Historically it was through the detection of a supernova in the Andromeda nebula that enabled American astronomer Heber Curtis to deduce, in 1917, that it must reside outside of the Milky Way Galaxy. Indeed, Curtis's observations eventually lead to the realization that the universe is composed of many billions of separate island galaxies.

An Expanding Universe

Once it became possible to determine the distances to nearby galaxies, astronomers started to map out their spatial distribution within the universe. When, however, the distance estimates were combined with additional observations relating to their speed, a startling result emerged, and it was realized that almost all galaxies had spectra that were red-shifted, indicating that they must be moving away from us.

First in 1918 and then again in 1924 German astronomer Carl Wirtz published research papers indicating that there was a systematic variation in the manner in which the galaxies appeared to be moving away from us. Indeed, Wirtz noted that his observations suggested a remarkable expansion of the system of nebulae, indicating that they were all rushing away from us.

The young American Astronomer Edwin Powell Hubble picked up on Wirtz's idea, and by 1929 he had amassed distance and velocity data (most of the latter being obtained by Milton Humason) on some twenty-four distant galaxies. With this data he published a graph showing recession velocity versus distance, and although the scatter was large, he was able to draw a reasonably good straight line through the data points. The result was electrifying – all galaxies beyond the Local Group of galaxies were moving away from us with a speed that increased systematically with distance. The data presented by Hubble indicated that the universe was truly expanding. Not only this, the data indicated that the universe was expanding uniformly, so that the recession velocity V of a galaxy is related to its distance D through an equation of the form $V = H_0 D$, where H_0 is Hubble's constant (as measured at the present epoch – hence the zero subscript). This simple formula tells us that the recession velocity increases linearly with distance and that the further a galaxy is away from us the faster it must be moving (Fig. 6.4).

Hubble's constant is one of the fundamental constants of cosmology, but it has taken the best part of 80 years for astronomers to agree upon its actual value. In his

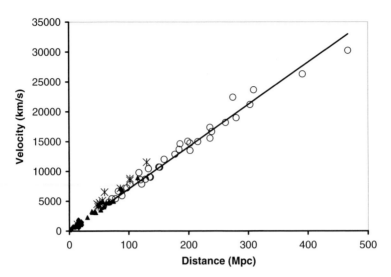

Fig. 6.4 Hubble's law is revealed in a plot of galaxy recession velocity against distance. The slope of the *diagonal line* corresponds to Hubble's constant H_0. The data shown is taken from the Hubble Space Telescope key project to measure Hubble's constant. The *circles* indicate data points derived from Type I supernova observations, the *triangles* relate to data points from spiral galaxy measurements, while the *crosses* indicate studies of elliptical galaxies. The slope of the *diagonal line* gives Hubble's constant as $H_0 = 72$ km/s/Mpc

1929 publication Hubble suggested a value of 500 km/s/Mpc for H_0, but modern-day estimates find it to be 72 km/s/Mpc. Although called a constant, Hubble's constant actually varies with the age of the universe; that this must be so becomes evident when we look more carefully at the units through which Hubble's constant is expressed – kilometers per second per megaparsec. Since both the kilometer and megaparsec are distance units they effectively cancel each other out (apart from a numerical conversion factor), and we are left with just the "per second" part. In these proper units, $H_0 = 2.33 \times 10^{-18}$ per second. The interesting feature of this number is that it provides us with an estimate for the age of the universe. Indeed, the approximate age of the universe is $T_{universe} \sim 1/H_0 \approx 13.6$ billion years. The critical density of the universe (discussed in the last chapter) also, it turns out, depends upon the value of Hubble's constant. Change the value of Hubble's constant, and you literally change everything.

Hubble's law tells us that the universe is expanding, and to cut to the core of the matter, the present consensus among astronomers is that not only is the universe expanding, but it is expanding at an accelerating rate, and it is for the latter reason that dark energy has been brought into play: literally, some energy must be driving the acceleration onwards. To see how this current dark energy-driven cosmological model comes about, we must first take a closer look at the brightest of astronomical standard candles, the supernova.

Death Throes and Distance

Any object with a known luminosity can be used to determine distance, as we saw earlier with the case of stellar candles. The brightest star that astronomers currently know of within our galaxy is the luminous blue variable (LBV) 1806-20. Located on the far side of the galaxy, some 12,000 parsecs away, this behemoth of a star is believed to weigh in at well over 100 times the mass of our Sun. Concomitant to its large mass, LBV 1806-20 has an equally impressive luminosity, which is estimated to be about 5 million times greater than that of sol. At the limit of human eye perception, under perfect viewing conditions and with no extinction effects, this stellar colossus would theoretically be visible if situated 16,000 parsecs away.

To a large, modern-day telescope working to a limiting stellar magnitude of say +20 (which corresponds to a brightness 40,000 times fainter than the human eye can record), the light from LBV 1806-20 would be discernable out to distances of order 10 million parsecs. At 10 Mpc we are certainly outside of the Local Group of galaxies but only half way to the Virgo cluster of galaxies. To probe the very boundaries of the observable universe we are going to need an even more luminous standard candle than our record-holding, simmering giant of a star, LBV 1806-20.

To measure distances beyond those provided by the brightest stars, we must, in fact, wait for them to die – for, indeed, it is by recording the incredible brilliance of their dying light that astronomers can map out the universe (Fig. 6.5). No longer encapsulated under the banner of standard candles, such supernova events might be better thought of as standard bombs, since the host stars generating them have literally blown themselves apart. There is a sting within the supernova tail, however,

Fig. 6.5 Supernova 1994d (visible to the *lower left*) outshines all of its stellar companions in galaxy NGC 4526. Situated within the Virgo cluster of galaxies (some 17 Mpc away) the supernova almost outshines the core of its host galaxy. (Image courtesy of NASA/ESA)

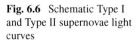

Fig. 6.6 Schematic Type I and Type II supernovae light curves

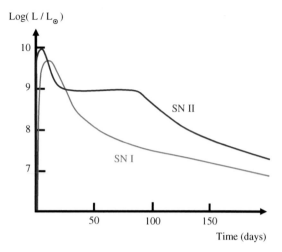

and it appears that not all standard bombs are born equal, and astronomers must distinguish between different types.

Just as zoologists distinguish between different kinds of animals according to their skeletal proportions, so astronomers distinguish between supernovae according to the shape of their light curves – plots of brightness versus time. Figure 6.6 illustrates the two main light curve types recognized, and they are appropriately called Type I and Type II. As the figure illustrates, the Type II supernovae are slightly more luminous at maximum brightness than Type I supernovae, and they also commonly display a brightness plateau. In contrast, the Type I light curves tend to show a steadier and more rapid decline in brightness with time.

Various additional supernova sub-types are also recognized by astronomers, but we need not consider these in detail. Just as there are two main supernova types, so there are two main mechanisms for their production. Type I supernovae are generated through the gravitational collapse of a white dwarf, while Type II supernova are the result of core collapse within massive stars.

The difference in the progenitor stars is crucial to understanding not only the light curve profiles but also the observed spectra and the galactic locations where the two supernovae types are observed. Although Type II supernovae are distinctly variable according to their maximum brightness, the Type I supernovae have relatively constant maxima, and accordingly it is the Type I supernova that are the most important objects with respect to cosmological distance measurements. We shall pick up on the story of Type II supernovae in the next chapter and for the moment concentrate on just the Type I events.

Future Sun – Take One

The Sun will eventually become a white dwarf, but it won't become a Type I supernova. The reason for this is simply because the Sun is an only child. If the Sun had formed in a binary system – and, in fact, most stars do form in binary systems – then its deep-future evolution would play out very differently. The Sun is presently 4.56 billion years young, making it about middle-aged for a main sequence star of one solar mass. Some 5–6 billion years from now the Sun will run into an energy crisis, and just as old astronomers waistlines tend to bloat out with increasing age, so the Sun will become a puffed-up red giant. The crisis that precipitates the Sun's red giant expansion is the exhaustion of hydrogen within its central core.

Throughout the Sun's main sequence phase it is the proton–proton chain of fusion reactions that converts four protons into energy and helium (the full reaction being $4P \Rightarrow {}^4He + 2e^+ + 2 \nu + energy$). It is the energy produced in this process that enables the Sun to remain in hydrostatic equilibrium and that replenishes the radiation lost into space at its surface – the same energy that has kept Earth warm and habitable for most of the past 4.56 billion years. The Sun is certainly massive, weighing in at some 2×10^{30} kg, or equivalently we might say that the Sun contains approximately 10^{57} protons. In spite of this vast number of protons within its interior, the Sun can only tap the proton–proton energy-generating chain for about 10 billion years.

After this time of peaceful, main sequence existence the innermost fuel supply runs dry, and the Sun must search for a new fuel to burn. This search, however, results in tumultuous structural changes, and the Sun must metamorphose. The change, however, is no transformation from a bland caterpillar to beautiful butterfly. It emerges from the main sequence as a roaring, bloated behemoth, a fire drake with deathly intent. Swelling perhaps to 150 times its present size, the Sun will consume, in a fearsome blaze, Mercury, Venus, and probably Earth. The destruction of the inner planetary system occurs as the Sun ascends the red giant branch in the Hertzsprung–Russell diagram (Fig. 6.7) – growing ever larger, cooler, and more luminous on the outside. Its inner core, in contrast, contracts and becomes denser and hotter.

Eventually, in a flash of explosive inner anger (point 4), the Sun finds its second wind, and a stable core helium burning phase begins. The aging Sun is now powered by the triple-α reaction in which three helium nuclei are converted into carbon and energy (the reaction runs as: ${}^4He + {}^4He + {}^4He \Rightarrow {}^{12}C + energy$). A time of luminous stability now holds (point 5 in Fig. 6.7), but it can't last forever since, once again, the supply of ${}^4He$ fuel is finite, and eventually it will grow scarce within the Sun's inner core.

The Sun is now dying, and it begins to feed upon the last vestiges of the hydrogen and helium that remain within its deep interior. Regions of nuclear burning flicker into existence around a degenerate carbon core, and the Sun swells up once again (point 6), becoming ever larger in size, with its core and outer envelope primed to undergo a final separation and terminal divorce.

Upon ascending what astronomers call the asymptotic giant branch the Sun begins to experience a series of thermal pulses, its size and luminosity alternately increasing then decreasing – a slow breathing cycle of last gasps, the stellar death rattle, prior to the core and envelope finally parting company. At this stage the Sun will become what is known as a long-period Mira variable. Eventually a furious wind develops at the surface, and the tatters of the Sun's outer envelope are driven off into interstellar space, eventually forming the great death blossom of a planetary

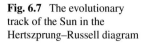

Fig. 6.7 The evolutionary track of the Sun in the Hertszprung–Russell diagram

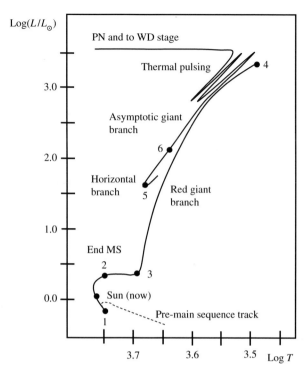

nebula (Fig. 6.8). And, at the core of this future nebula resides the Sun's dormant core – now transformed into a diminutive, incredibly hot, low luminosity white dwarf.

The future white dwarf Sun will slowly cool off, radiating its residual thermal energy into the coldness of space. Eventually it will cool down to the temperature of the microwave background, and unless some exotic and presently unknown nuclear decay process begins to operate, the eventual black dwarf Sun is fixed in place forever – a frigid, zero luminosity, Earth-sized, half-solar mass conglomeration of carbon-rich matter supported by the degenerate, non-overcrowding pressure of electrons – a veritable diamond in the sky.

The Degenerate World of White Dwarfs

White dwarfs are extreme objects with incredibly high interior densities. Detailed computer models indicate that the white dwarf that will eventually form as the Sun undergoes its terminal planetary nebula phase will have a mass of about 0.6 solar masses (the other 0.4 $M_\odot$ of its original mass being ejected into interstellar space by strong surface winds). With a size similar to that of Earth (that is, about 100 times smaller than the present-day Sun), the bulk density of the future white dwarf Sun

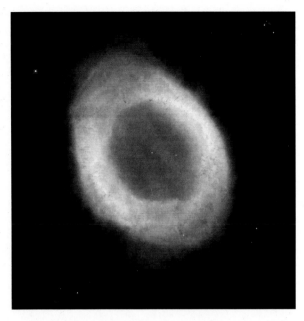

Fig. 6.8 The Ring Nebula (M57), a planetary nebula located in the constellation of Lyra. UV photons emanating from the hot white dwarf at its center cause the ionization of the surrounding hydrogen cloud via the reaction UV + H $\Rightarrow$ P + e$^-$, creating a sea of protons and free electrons. In the outer regions of the ionized cloud the reverse process takes place, and protons recombine with electrons to produce excited hydrogen atoms. The excited electron eventually quantum jumps to a low energy orbit, and an emitted photon carries away any excess energy (schematically, P + e$^-$ $\Rightarrow$ H + photon). It is the release of myriad photons in the recombination process that paints the visual display of the planetary nebula. M57 is some 700 pc away. (Image courtesy of NASA)

will be of order 2 billion kilogram per cubic meter. Indeed, a 1-cm sided cube of typical white dwarf matter weighs in at a staggering 2,000 kg (two metric tons). The closest white dwarf to Earth is Sirius B (Fig. 6.9), the binary companion to Sirius A – the brightest star in our sky. Situated just 2.64 pc away, the light from Sirius B is only just visible to us, its brighter companion Sirius A being of order 1,000 times more luminous, almost completely obliterating its fainter companion from view.

A solar mass white dwarf, as exemplified by Sirius B, has a bulk density that is some 1.5 million times greater than that of the present-day Sun. Having such properties the interior of a white dwarf cannot be treated as behaving like a perfect gas, where there are minimal interactions between the constituent particles. Indeed, within white dwarf interiors the matter is degenerate, and accordingly quantum mechanical effects come into play. Specifically, the electrons are so closely packed together that the Pauli Exclusion Principle comes into action, and a non-overcrowding pressure develops. This remarkable state of affairs dictates that white

Fig. 6.9 Sirius A and B. The white dwarf Sirius B is seen in this image as the fainter "dot" to the *lower left*. Sirius B is about 12,000 km across, about 38 times less luminous than the Sun but weighs in with a mass of 0.98 $M_\odot$. (Image courtesy of NASA/HST)

dwarfs remain, at least in principle, forever stable, since the outward-acting electron degeneracy pressure can hold gravity in check in perpetuity irrespective of the interior temperature.

The first detailed quantum mechanical description of white dwarf interiors was developed during the 1930s by the then young and future Nobel Prize winning physicist, Subrahmanyan Chandrasekhar (Fig. 6.10). Born in India, but later moving to England and then eventually to the United States (University of Chicago), Chandrasekhar was a remarkable theorist who, over his lifetime, wrote definitive research papers and textbooks on stellar structure, stellar dynamics, fluid mechanics, radiative transfer, general relativity, gravitational waves, and black holes. With respect to the story of white dwarfs, however, Chandrasekhar discovered that they are only stable up to a specific limiting mass, appropriately called the Chandrasekhar limiting mass M_C.

The limiting mass for white dwarfs comes about as a result of the further squeezing of the electrons. A remarkable feature of white dwarfs is that they become smaller with increasing mass. Adding matter to a white dwarf, therefore, makes it shrink. It is this additional shrinkage with increasing mass that ultimately sets a white dwarf on the road to gravitational collapse and ruin.

Provided that the typical velocities of the electrons remain much less than that of the speed of light, then a white dwarf can remain stable no matter what its internal temperature. If, on the other hand, the electrons begin to move with relativistic speeds, as a result of mass accretion squeezing the white dwarf to a smaller size, then

Fig. 6.10 Subrahmanyan
Chandrasekhar (1910–1995),
one of the great founding
fathers of stellar astrophysics,
and the first person to
describe white dwarf
structure in terms of a
detailed quantum mechanical
framework

a remarkable change comes about in the internal pressure law. While the pressure is still independent of the temperature, a critical mass develops beyond which no stable white dwarf configuration is possible. Detailed numerical calculations indicate that the limiting mass is $M_C \approx 1.44$ M$_\odot$, and once a white dwarf approaches and then exceeds this mass it will become unstable and gravitational collapse must occur; there are no exceptions to this rule, and there are no survivors from any transgressions. This strong limit to the maximum mass of any white dwarf is a key point with respect to explaining why Type I supernovae are standard bombs. The key point is that they all detonate upon attaining the same mass – the mass limit being M_C.

Future Sun – Take Two

Let us travel on a flight of fancy. Imagine that the Sun formed within a binary system. The exact system details are not our main concern here, but let us assume that the Sun has a slightly less massive, relatively close-in partner. (We will also be virtual observers in this system and are not specifically concerned whether planets will have actually formed around our doppelganger Sun.)

The mass difference that exists between the two stars dictates that the congruous Sun will become a white dwarf before its partner does – lower mass stars evolve more slowly because of their reduced intrinsic luminosities. Up to this point our doppelganger Sun has evolved in exactly the same way as described earlier. We eventually end up therefore with a binary system in which our imagined Sun has become a white dwarf while its (lower mass) companion is still a main sequence star generating energy within its core by converting hydrogen into helium.

Eventually, however, the lower mass companion must evolve into a red giant, and accordingly it will swell up to gigantic proportions. The key point at this stage is that if our two make-believe stars are close enough to each other, then the white dwarf Sun will begin to move within the bloated envelope of its companion and thereby accrete matter (Fig. 6.11). The mass of the white dwarf Sun steadily grows, and accordingly its mass creeps ever more closely towards the Chandrasekhar limiting mass. In this scenario our imagined white dwarf Sun has no quiescent dotage, and it eventually accretes sufficient matter that the electrons within its interior are forced to move with higher and higher velocities – eventually becoming relativistic. At this stage gravitational collapse is inevitable. With the catastrophic collapse of our imagined white dwarf Sun a Type I supernova results, and our "take two" Sun is no more (*draco dormiens nunquam titillandus*).

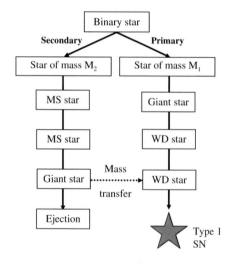

Fig. 6.11 One possible sequence of events leading to the production of a Type I supernova in a binary star system. Time is imagined to be increasing downward in the diagram. The initially more massive primary star evolves into a white dwarf first and will eventually begin to accrete matter from its aging (secondary) companion. Finally, the white dwarf is pushed over the Chandrasekhar limit and collapses to produce a supernova

Although the scenario for the binary-born, doppelganger Sun just described is imaginary, it will play out for real within other genuine multiple-star systems located within our galaxy (and indeed, within other galaxies). All Type I supernovae begin from the same starting point – the gravitational collapse of a white dwarf upon reaching the Chandrasekhar limiting mass. They are indeed standard bombs. For cosmology this result is of extreme importance, since it tells us that whenever we see a Type I supernova (as identified by its light curve) then we know that it must have been produced through the collapse of a white dwarf, and that its peak brightness must correspond to that of a standard candle. Any difference, therefore, between the observed peak brightness and the standard candle brightness must be proportional to the distance between us and the supernova location.

The Case of IK Pegasus B

Type I supernovae are observed to occur within all galaxy types, and they must occasionally occur within our own Milky Way Galaxy. Since 1572, however, only two Type II supernovae have been visually observed. (The remnants of two other supernovae are known, but their detonation dates are uncertain, and they were not historically recorded). The first of the historic galactic supernovae was widely observed from across northern Europe during late 1572 (Fig. 6.12), and the great Tycho Brahe – the Noble Dane – produced a remarkable pamphlet concerning his thoughts on this new heavenly phenomena. The second galactic supernova was observed in October of 1604 and was described by Brahe's equally great contemporary Johannes Kepler.

Fig. 6.12 X-ray image of SN 1572. Observed throughout Europe this supernova was located in the direction of Cassiopeia and was triggered by the coalescence of two white dwarfs. (Image courtesy of NASA)

The most recent naked-eye supernova (a Type II supernova) visible from Earth detonated in the Large Magellanic Cloud in 1987, and while it was not actually seen, astronomers have very recently discovered the remnant of a supernova that occurred about 25,000 years ago and which technically should have been visible from Earth in the late 1860s (Fig. 6.13).

We do not know when the next galactic supernova will be seen with the unaided human eye, but one candidate system, known as IK Pegasus, has been detected that

Fig. 6.13 Supernova remnant G1.9+0.3. This composite radio and X-ray wavelength image shows an expanding bubble of hot gas pushing into the surrounding interstellar medium. This supernova wasn't observed visually since it is located close to the center of our galaxy – a direction in which there is heavy dust obscuration. (Image courtesy of NASA)

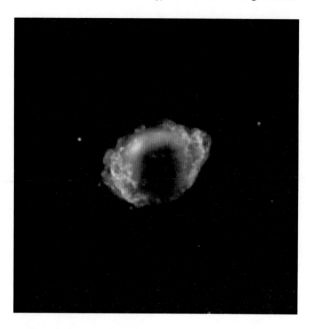

will sooner or later produce a Type I supernova in our interstellar backyard – in fact, it might be an uncomfortably close supernova.

IK Pegasus is a binary star system located about 46 light years from the Sun. It is composed of a white dwarf and a relatively massive, hot, and luminous so-called A-type star companion. The two stars orbit the system's center of mass once every 22 days and are about 39 $R_\odot$ apart.

To give some idea of the scale, the two stars in IK Pegasus are two times closer together than Mercury's orbit about our Sun. With an estimated mass of 1.15 $M_\odot$, IK Pegasus B is one of the most massive white dwarfs known, and it is therefore relatively close to the Chandrasekhar limit. The primary star of the system IK Pegasus A is presently in its main sequence phase, but it must eventually puff up into a giant star, and from that moment on IK Pegasus B is under sentence of death. Once IK Pegasus B has accreted an additional 0.29 $M_\odot$ of material from its giant companion it will be pushed over the Chandrasekhar limit, resulting in its collapse and the production of a Type I supernova.

We cannot tell for certain when IK Pegasus B will explode – it depends upon the accretion rate of material from its companion, and this rate is currently unknown. When it detonates, however, it could well be much closer to the Sun than it is at present. The system is now heading towards us with a space velocity of some 20 km/s and will be at its closest approach in about a million years from now. If IK Pegasus B undergoes collapse at this time the system will be 37 pc away from us. At this distance supernova IK Pegasus will achieve a peak brightness in excess of that displayed by the full Moon – a bedazzling celestial display indeed.

High-Z Supernova Surveys

The extreme brightness of supernova make them ideal standard candles for probing the distances to galaxies, and although they may not be observed very often within our Milky Way Galaxy, they are continuously observable in the multitude of other galaxies that exist within the observable universe (Fig. 6.14). Recall from Chapter 1 that there are an estimated 100 billion galaxies in the observable universe. By patrolling the skies for the transient light shows of Type I supernova the distances to remote galaxies can be measured, and by combining these distances with recession velocity data a map of the Hubble flow can be obtained.

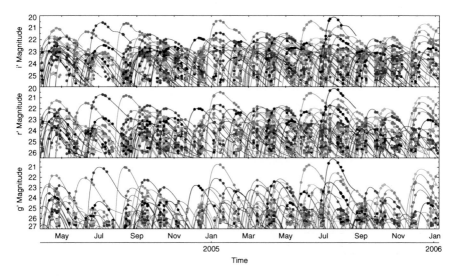

Fig. 6.14 Supernova legacy survey results. This image shows the *light curves* (each recorded through three different filters) of more than 150 Type I supernova discovered by astronomers using the Canada-France-Hawaii Telescope. The multiple wavelength measurements allow astronomers to make a better estimate of each supernova's maximum brightness and type. (Image courtesy of M. Sullivan)

The expectation from the standard Big Bang model is that the expansion rate of the universe should slow down with age. Remarkably, however, evidence first gathered in the early to mid-1990s appears to indicate that the exact opposite effect may be happening – rather than the universal expansion slowing, it appears to be accelerating.

What exactly is it that the high–Z survey results show? Well, first of all we should explain what the Z in high-Z means. The Z term is actually shorthand for the wavelength stretching of light due to the expansion of the universe. If we picture a light ray as an oscillating wave, then its wavelength corresponds to the spatial distance over which the wave profile repeats itself – in the case of light, this is a distance of about 10^{-7} m.

Now, given that the light emitted from a distant galaxy, in the distant past, has a wavelength λ_0, then in the time interval that it takes the light ray to reach our telescopes on Earth the universe will have expanded, and the wavelength of the light will have been stretched out to a value λ. The Z term is a measure of the light stretching (or cosmological red-shift effect) that has taken place during the light ray travel time, and it is defined by the expression $1 + Z = \lambda_0/\lambda$.

So if, for example, the universe has expanded by a factor of 2 during the time interval that a light ray was emitted from a galaxy and its being detected by us, then the wavelength of the light ray will have also been stretched out by a factor of 2 (i.e., $\lambda = 2\,\lambda_0$), and accordingly $Z = 1$; the larger the value of Z, the smaller the size of the universe at the time when the light was emitted. Using this effect to its full advantage, the idea behind the high-Z supernova surveys is to look for Type I supernovae in galaxies with various values of Z. Since the size of the universe is encoded within the (measurable) Z value, by comparing the brightness of the supernova at different Z values (i.e., at $Z = 2$ and $Z = 0.1$), so the relative size and expansion rate of the universe can be deduced at different epochs.

Dark Energy and ΛCDM Cosmology

As is often the way in science history, two independently working research groups announced the discovery of an accelerating universe at almost exactly the same time in 1998. Both groups, one led by Saul Perlmutter (Berkeley National Laboratory), the other by Brian Schmidt (Australian National University), were using studies of Type I supernovae to determine the variations of their peak brightnesses with Z. The arguments advanced by the two research groups indicated that the most distant (high Z) Type I supernovae were fainter than expected, that is, the light had traveled further than anticipated according to the expansion rate of the universe deduced from the supernovae observed in nearby (low Z) galaxies. Rather than the universal expansion slowing down, or even coasting at a constant velocity, it appears to be increasing. We apparently live in a universe that will grow larger and larger at an ever-increasing rate. Rather than experiencing a terminal Big Crunch, the universe may be destined to undergo a Big Rip.

An accelerating universe must have an energy source – something has to drive the acceleration – and this is where dark energy comes in. Given that the acceleration is a real phenomenon, then we are left with the conclusion that dark energy must permeate all space but does not interact with any type of matter, whether baryonic or dark. Since Einstein's mass-energy equivalency formula allows us to express the dark energy in terms of a mass density, and current estimates suggest that it is of order 10^{-26} kg/m^3. (This is equivalent to the density of six hydrogen atoms within a cube having 1-m sides.) Every region of space – the Local Group of galaxies, the Milky Way Galaxy, the Solar System, your living room – is permeated by this amount of dark energy. It is everywhere.

At the beginning of this chapter it was indicated that dark energy accounts for about 73% of the matter density within the universe (Fig. 6.1). This estimate is derived upon the experimental results that indicate we live in a flat universe with $\Omega = 1$ (as posited by the Big Bang inflation model and observed from the structure of the microwave background). Since the contributions to Ω from baryonic and dark matter add up to $\Omega_B + \Omega_{DM} = 0.27$, so the contribution from dark energy must be $\Omega_{DE} = 1 - \Omega_B - \Omega_{DM} = 0.73$. This calculation presupposes, of course, that there are not additional items contributing to the matter density of the universe – contributions that we are presently completely unaware of. Figure 6.15 indicates the range of possibilities.

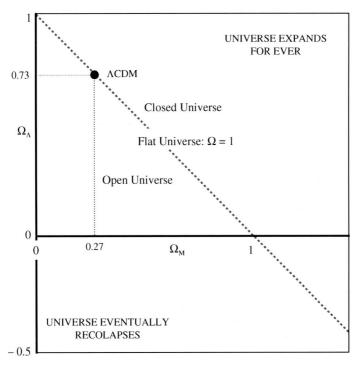

Fig. 6.15 Constraints on the cosmological densities for dark energy and matter (both baryonic and dark) in a flat universe. Observations of the cosmic microwave background indicate we live in a flat universe

In terms of the standard cosmological model the present best description of the universe's dominant features are that it is an expanding and accelerating universe that contains some predominant form of clumped, cold dark matter that came into existence approximately 14 billion years ago. When just 10^{-35} s old the universe underwent a transition that resulted in a rapid inflation in its size. This process fixed the spatial geometry to be Euclidean (that is, flat, with $\Omega = 1$). The initial universe was exquisitely hot, incredibly dense, and radiation dominated, with baryonic matter only beginning to appear after the first few minutes of expansion. After some 7–8

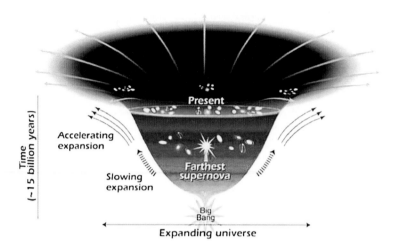

Fig. 6.16 The growth of the universe. Following an initial expanding but slowly decelerating phase, the repulsive effects of dark energy began to take hold and thereafter drive an accelerating expansion. (Image courtesy of NASA)

million years of gradually slowing expansion, the influences of dark energy began to grow in importance, resulting in the accelerating universe that the Type I supernovae apparently reveal to us today (Fig. 6.16). Only at the very tip of the great accumulation of matter and energy within the universe (at a 4% contribution level) are the visible stars, galaxies, clusters of galaxies, and intergalactic gas.

The present-day best-fit to the observations ΛCDM cosmological model is incredible, elegant, and stranger than fiction. It accounts for the available data, but it leaves astronomers and physicists with the somewhat unnerving problem that 96% of the matter content of the universe is in a form that has absolutely no independent experimental understanding and/or verification – yet. Not only this, since there is no consistent and/or agreed upon theory to explain why dark energy should even come about, it is not clear if the universe will continue to accelerate forever, of if it will begin to slow down and then rapidly decelerate a year, a million years, or a trillion years from now.

Researchers will be spending many years, if not many decades, pouring over the vast mountain of data that the LHC is going to generate. New particles are almost certainly going to be found, and the Standard Model will, with little doubt, see revisions as a result of new understandings. This being said, there is no specific expectation that the LHC experiment will yield any direct insight on, or evidence for, dark energy. Should the existence of dark energy be established beyond any reasonable doubt, however, then it is clear that new physics must exist, and this is where the LHC experiment might well enable theorists to distinguish between rival theories. Some supersymmetric theories, for example, explicitly require the cosmological constant to be zero, and accordingly if Λ is found to be any other value than that, then at least those theories can be retired.

The assumption that all Type I supernovae have the same peak luminosity is central to the dark energy hypothesis, and while there is good reason to believe that this assumption is basically true, a few worrying signs concerning its robustness are beginning to emerge. Indeed, writing in the August 13, 2009, issue of *Nature*, Daniel Kasen (University of California, Santa Cruz) and co-workers describe a series of detailed computer models simulating Type I explosions. Importantly, the researchers find that the explosions are likely to be asymmetric, occurring off-center and closer to the surface than had previously been realized. This asymmetry means that the measured peak brightness can vary by as much as 20%, according to whether the emerging blast-wave is directed towards the observer or away.

It is also possible that changes in chemical composition might systematically affect the peak brightness, too. This could mean that there is a variation in peak brightness according to how old the universe was at the time when a particular supernova's parent binary-system formed. Much yet remains to be studied with respect to the Type I supernova models, and until we are certain that the constant peak brightness assumption is genuinely true, then there will continue to be doubt about the accelerating universe model and its implied existence of dark energy.

Some theorists find the implications of dark energy so extreme that they have begun to explore alternative explanations for the apparent universal acceleration. It is entirely possible, for example, that Einstein's theory of general relativity, the mathematical backbone of modern cosmology, requires revision on a scale comparable to that of galaxy clusters and larger. What these revisions are, and how they might be experimentally verified, are presently unclear, and while the MOND model offers one potential solution to the dark matter problem (in the sense of saying that there is no dark matter), it provides no explanation for dark energy.

One radical approach to solving both the dark matter and the dark energy problems has been proposed by Hong Sheng Zhao of St. Andrews University, Scotland. Zhou suggests, in fact, that they are a manifestation of the same phenomena that operates differently on different physical scales. Indeed, Zhao advocates the existence of what he calls a chameleon-like dark fluid (no less mysterious in its origin perhaps than dark matter or dark energy) that permeates all of space. On the galactic scale Zhao's dark fluid behaves like dark matter, but on larger scales it behaves more like dark energy. In this model dark matter, for example, is not thought of as an elementary particle but is rather considered to be a packet of energy, or dark fluid that propagates through space. Accordingly Zhao makes the clear prediction that the LHC experiment will find no evidence for dark matter particles – a bold prediction, for sure, but one that is very much in resonance with the way in which science really works; make a prediction and see if it holds true.

An even more radical approach to solving the dark matter and dark energy problems, again in the sense of arguing that they are in fact an illusion, has been proposed by Jose Senovilla and colleagues working at the University of the Basque Country in Bilbao, Spain. This work takes its lead from the multi-dimensional aspect of superstring theory and suggests that the phenomena we presently call dark matter and dark energy can be explained away by allowing time to slow down as the universe ages. Indeed, the argument presented by Senovilla and his colleagues makes the prediction that what we currently consider to be the time dimension of four-dimensional spacetime is slowly turning into a spatial dimension. In this model, billions of years into the future, time (as we know it) will eventually end and thereafter cease to be of any relevance in the workings of the universe. If it doesn't actually stop beforehand, time will presumably tell if this theory is on the right track!

Lest we might forget, we live in a wonderful age of opportunity, and it seems fair to say that the apparent existence of dark energy will provide for the discovery of new physics and the development of new physical theories. One is nonetheless left, however, with the uncomfortable feeling that the words of pioneer geneticist and biologist J. B. S. Haldane are beginning to ring more and more true: "The universe is not only queerer than we suppose, but queerer than we *can* suppose."

A Distant Darkness

If the current ΛCDM cosmological model is correct, then the universe will expand forever at an ever-increasing pace. It will never re-collapse to undergo a Big Crunch, and the deep future promises to reveal a far less crowded universe than we presently see. Indeed, only gravitationally bound structures, such as the Local Group of galaxies (Fig. 6.17), will survive universal expansion. All else will eventually disappear beyond the observable horizon.

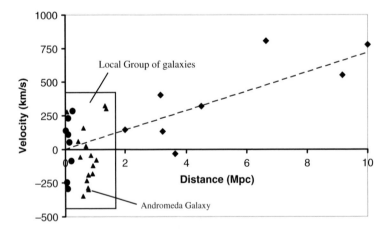

Fig. 6.17 The Hubble diagram for galaxies within 10 Mpc of the Milky Way. Within the Local Group the gravitational interaction between galaxies can override universal expansion – allowing for orbital rather than linear expansion motion. Only beyond about 5 Mpc does the universal expansion begin to dominate and Hubble's law (shown by the *dashed diagonal line*) hold true. *Triangles* are Local Group galaxies, *circles* are satellite galaxies to the Milky Way, and *diamonds* are nearby, but not Local Group, galaxies

Indeed, the universe of the deep future promises to be a much easier one to understand, with its entire baryonic content being composed of just one metagalaxy – the collisional remnant of what was once the Local Group of galaxies. Outside of the metagalaxy there will be a smattering of dark matter and a dominance of dark energy. And even this bleakness will eventually simplify further. The universe will evolve to an all-encompassing stygian darkness. Eventually there will be no stars at all, and the entire "visible" universe will consist entirely of stellar remnants (black dwarfs, neutron stars, frozen brown dwarfs, planets, and black holes), a smattering of hydrogen and helium atoms, dark matter, and a dominance of dark energy.

Our universe, in fact, appears to be going through a process of erasing its past. As the cosmic acceleration continues deep into the distant future, so the possibility of observers deducing that it had a Big Bang, a hot and dense beginning, will fade. Due to the Z-stretching effect the microwave background photons will be red-shifted deep into the very long radio wavelength portion of the electromagnetic spectrum, and the continued cycling of hydrogen through star formation and stellar evolution will push the cosmic abundance distribution ever further away from that produced within first few minutes of primordial nucleosynthesis.

Physicist Fred Adams (University of Michigan) has rightly described the present-day universe as being in a stelliferous phase. Stars, the bright cauldrons of baryonic matter, are all around us, and they are slowly changing the characteristic makeup of the universe. Astronomers estimate that the interstellar medium accounts for about 5–10% of the mass of our galaxy. This mass, however, translates into at least 10 billion times the mass of our Sun, but as with any finite resource, no matter how large and inexhaustible it might seem at first, without the complete recycling of material the process of star formation must eventually come to an end.

Indeed, an epoch will eventually arrive, some 100 trillion years from the present, when our galaxy will consist entirely of a final generation of very low mass stars, white dwarf remnants (slowly cooling to become black dwarfs), neutron stars supported by degenerate neutron pressure, brown dwarfs, planets, and black holes. None of these latter objects have significant internal energy-generation sources, and even if initially relatively bright, they will eventually fade to complete and absolute blackness. Ultimately, after countless eons of rambling motion, even these remnants will be mostly absorbed through collisions with massive black holes. And those degenerate remnants that are fortunate enough to avoid black hole spaghettification, will not survive beyond the lifetime of the proton.

It is not known for certain if the proton is eternal, but some theories suggest that it has a decay time of perhaps 10^{40} years. Current experimental estimates give the decay time as being greater than at least 10^{33} years (see Chapter 7). Once this distant epoch is realized the fate of any surviving black dwarfs and neutron stars is sealed, and they will gradually evaporate their remaining mass-energy into space. Eventually the enveloping darkness of the universe will be almost complete. As the oppressive black hole epoch begins to unfold, the entire visible universe, supposing that there are living entities capable of seeing it, will be illuminated entirely by the faint glow of Hawking radiation (see Chapter 7). It would appear that T. S. Elliot had (without actually realizing it) the right image in mind when he wrote in his poem, *The Hollow Man*:

This is the way the world ends
Not with a bang, but a whimper.

Testing Copernicus

Nearly all present-day cosmological models are constructed according to the premise of the cosmological principle. Indeed, universal expansion (Hubble's law) is a direct consequence of the large-scale homogeneity of the universe. But how sure are we of the correctness of the cosmological principle? This is a rather radical question, since the truth of the principle cuts to the very core of modern cosmology. If the cosmological principle is wrong, then we can never hope to fully understand the entire history of the universe. Indeed, its true origin and its deep future would be entirely lost to us.

That one might question such deeply set principles is perhaps rather surprising, but a growing number of astronomers and physicists are asking this very question in light of the equally extreme and unusual properties associated with dark energy. Is it possible, one might ask, that dark energy is really just an illusion brought about by the fact that we (that is, the Local Group) occupy a very special place within the universe? Such radical thinking goes against the grain of another principle upon which all modern cosmology is built – the so-called Copernican principle.

It was Copernicus that displaced Earth and its human cargo away from the center of the medieval universe. Scientists in the modern era have raised Copernicus's action to the level of a self-evident truth; indeed, it is a principle upon which all work proceeds, and accordingly it is implicitly assumed that as observers we do not occupy any special location within the universe. The sheer incomprehensibility of dark energy, however, has caused some physicists to ask if it might not be a kind of mirage or manifestation of the fact that our galaxy is actually located at the very center of a spherically symmetric, under-dense, bubble-like region that just happens to have a size comparable to that of the observable universe. Such a configuration, it has been argued, can explain the Type I supernovae data without invoking dark energy, but it comes at the rich price of saying that the Copernican principle is wrong and that as observers we do occupy a very special place within the universe.

It would be far too premature to suggest that the Copernican principle is crumbling in light of the presently available cosmological observations, but it is exciting to think that we are on the threshold of being able to decide upon its correctness. As more high-Z Type I supernovae data is gathered in, so the possibility of testing the truth (or falsity) of one of the highest held principles of science will be realized. Perhaps the next big step in this debate will come about through NASA's Joint Dark Energy Mission (JDEM), provisionally scheduled for launch in 2015, and ESA's Euclid Mission (see Chapter 5), slated for launch in 2017. The JDEM mission will be jointly sponsored by NASA and the US Department of Energy and will attempt to make precise measurements of the universe's expansion rate, looking specifically to determine how the rate might have changed as the universe has aged.

To say that there is much that we don't know about the universe is a profound understatement, but it is this very feature that makes cosmology such a challenging and exhilarating subject. What is perhaps the most exciting part of cosmology, however, is that – irrespective of where we think our theories might stand – the universe always manages to find a new way of surprising us.

Chapter 7
The Waiting Game

There is no excellent beauty
that hath not some strangeness
in the proportion.

Francis Bacon (1561–1626)

Particle physics experiments, by their very nature, produce statistical results. Unlike flicking the switch on a radio or a wall socket where there is an instantaneous response – a light goes on or a sound is produced. When the LHC is in operation any number of several billion possible things can happen per second. Most of the 2×10^{14} collisions that will occur each and every day that the LHCs is in full operation will be of no interest to the CERN researchers – hence the complex event screening described in Chapter 3. What were once new and triumphant signals obtained on older colliders will be just background noise in the new experimental detectors, and the investigation of such standard, well understood events is not the LHC's primary goal. Today's researchers have more exotic quarry in their collective sights, but they will have to sieve through their colossal pile of selected data with exquisite care.

Hoping for the Unexpected

Ask any scientist, and he or she will tell you that in addition to the sheer joy of studying the chosen subject, science is all about testing, checking, and developing theories to explain experimental data, and then testing and checking the theories once again. It is this cyclical nature of science that makes it such a powerful tool; make predictions, do an experiment, and see if the predictions are correct. If the theory gives the wrong predictions, then adjust it (if possible), else throw it out and start over afresh.

Even the most cherished theory must be rejected if it provides incorrect predictions. There is no safe haven for half-truths in the world of science. Indeed, science has no mercy, and incorrect ideas and faulty explanations will not be tolerated for very long. At the heart of science, as indicated earlier, is experimental testing. Theories literally live or die (at least ultimately) by experimentation, and for

M. Beech, *The Large Hadron Collider*, DOI 10.1007/978-1-4419-5668-2_7,
© Springer Science+Business Media, LLC 2010

particle physics this is exactly what the LHC is all about. The many hard years of planning, design, re-design, development, construction, more re-designs, and testing have all been endured by thousands of researchers purely in order to test the deeper secrets of the Standard Model. The CERN researchers, through the highly sophisticated design of their experiments, know exactly what it is that they are looking for, and they know exactly what they expect the collider to produce. The question now is what will be seen. In 10 years from now all will be clear; the new results will have been gathered in, the LHC will have completed its primary function, and the predictions based upon the standard theory of particle physics will be either right or wrong. This is all to the good, and no physicist will argue with the importance of knowing the eventual outcome.

There is more, indeed very much more, however, to the workings of science than the direct experimental testing of new ideas and favored theories. Chance results, the revealed gifts of *fortuna*, and the entirely unexpected also play a major role in advancing our knowledge. Indeed, it is the detection of phenomenon that the favored theories *don't* predict that will drive science forward. It is with such thoughts in mind that the LHC researchers will be looking for the unexpected, events that have no clear explanation or are counter to expectation.

There are very high hopes that the LHC will reveal new and entirely unanticipated results, providing possible clues, as we have described in earlier chapters, to the properties of dark matter and dark energy; it might also provide us with clues as to which, if any, of the many beyond standard theory models are on the right track. The LHC may also, of course, produce a host of entirely unexpected and potentially dangerous phenomena (as discussed in Chapter 2).

Massive Star Evolution

Before we can understand the possible exotic objects that the LHC might produce and how we can be sure that they offer no serious threat to life on Earth, we must first go on a stellar journey. Indeed, we need to embark upon a cosmic odyssey in which the events of tens of millions of years of time are compressed into a few paragraphs and a brief survey of massive star evolution. Let us consider the life of a star 30 times more massive than our Sun – a stellar behemoth by any standards.

The life story of a 30-solar mass star can be summarized as, shine very brightly and die young. Indeed, within the same time that it takes our Sun to consume the hydrogen within its central core, a sequence of some 4,000 30-solar mass stars could run their entire evolutionary lives. Indeed, if the Sun is a shining beacon then a 30-solar mass star is a stellar firefly – here today and gone tomorrow.

Compared to our Sun, everything about the life of a massive star is hyper-accelerated. Although at first thought one might logically expect a massive star to have a much longer main sequence lifetime than our Sun – it has, after all, much more hydrogen fuel – it turns out that it is the enhanced luminosity of massive stars that dramatically shortens their lives. The indomitable British astrophysicist Sir Arthur Eddington first noted in the early twentieth century that a star's luminosity (L) is closely related to its mass (M). Indeed, the luminosity increases according to a power law with $L \sim M^{3.5}$. What this relationship reveals is that if we compare the luminosity of, say, a 2-solar mass star to that of the Sun, the 2-solar mass star will be about 11 times more luminous (i.e., $2^{3.5} = 11.314$) than the Sun.

Further, the luminosity of a 30-solar mass is about 150,000 times greater than that of the Sun. Massive stars shine very brightly, and it is because of this that they soon run into an energy crisis. If a star could convert all of its hydrogen into helium through nuclear fusion reactions, then Einstein's formula $E = Mc^2$ tells us what the total amount of energy E a star of mass M has to work with. Given that a star uses up its fuel supply at a rate determined by its luminosity, we can estimate the main sequence lifetime T_{MS} by taking the ratio $T_{MS} \approx$ (fuel available)/(rate at which fuel is consumed) $\approx E/L \approx Mc^2/L$. The reason for the short lifetime is now revealed, since if we combine the power law mass-luminosity relationship found by Eddington with the expression for T_{MS} we find that T_{MS} (M) $\approx T_{MS}(SUN)/M^{2.5}$, where M is expressed in solar mass units. In other words the longest lived hydrogen fusion phase of a 30-solar mass star will be about $1/(30^{2.5})$ times shorter than that of the Sun – this is about one 5,000th of the Sun's main sequence lifetime.

Massive stars by shining very brightly evolve rapidly towards a final stage (Fig. 7.1). Once all the hydrogen is consumed within their central cores, they can start to convert helium into carbon, and then once all the helium has been consumed

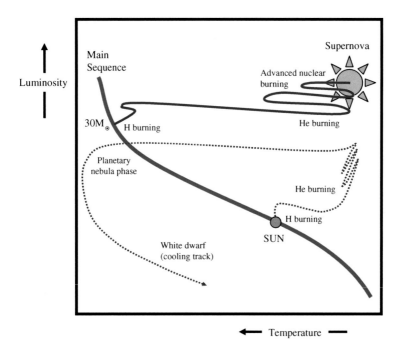

Fig. 7.1 The journey of a 30 M$_\odot$ star across the Hertzsprung–Russell diagram. The evolutionary track of the Sun is shown for comparison

they can generate energy by converting carbon into neon, and so on. At each fusion step following the main sequence phase, however, the returns become smaller and smaller. Each new fuel is less abundant than its predecessor, and accordingly the fusion reactions rapidly work their way towards the fusion of iron. It is with the generation of an iron core that the fate of a massive star is sealed. Energy cannot be extracted by fusion reactions acting upon iron, and accordingly something will have to give.

By the time that a massive star starts to build an iron core the central density will have climbed to a staggering compression of 10^{10} kg/m^3, and the central temperature will have reached about 5 billion degrees – as close a version to hell as anyone might imagine. And this hellish iron core is, perhaps appropriately, destroyed by the emergence of a brilliant light. Specifically, iron nuclei are destroyed through photodisintegration with gamma ray photons. The iron core, which at the time of its destruction has a mass about equal to that of our Sun and a size comparable to that of Earth, is catastrophically broken down by the gamma rays into a spray of helium nuclei and neutrons. Within the blink of an eye, a time encompassing just a second or so, the entire core is destroyed, and the central pressure support vanishes.

No longer able to support the weight of overlying layers, an explosive collapse of our massive star is now unavoidable. The journey of the core is not, however, ended, and incredibly it collapses down even further to form a neutron star – a sphere perhaps just 20–30 km across, with a density comparable to that of an atomic nucleus. It is the neutron star that will be of interest to our story later on.

The evolutionary journey between a 30-solar mass star and the Sun could hardly be more different. As Fig. 7.1 illustrates, they both begin and end their lives in completely different regions of the Hertzsprung–Russell (HR) diagram. The 30 M$_\odot$ star is initially some 100,000 times more luminous than our Sun, about 7 times larger, and nearly 7 times hotter. The Sun, as we saw in the last chapter, will eventually evolve into a white dwarf with a size comparable to that of Earth. In contrast, after consuming its hydrogen, a 30 M$_\odot$ star will evolve into a luminous red supergiant, where after a brief helium and advanced nuclear burning phase it will explode as a supernova (Fig. 7.2). The Sun and a 30 M$_\odot$ star end their days in opposite corners of the HR diagram – one winking out as a small and faint white dwarf, the other catastrophically blasting the greater amount of its initial mass back into the interstellar medium, and for a few brief days outshining all of the other stars within the galaxy.

Supernova explosions are among the most energetic events that can be observed within the universe. They are primal blasts, but embedded within the chaos of their destruction is the fusion necessary for the generation of all the elements beyond iron in the Periodic Table. Earth, all the other planets in the Solar System, and life itself are all made possible because myriad massive stars have run the course of their evolution and expired as supernova.

However, this incredible alchemy, described in Chapter 1, is not our main interest now. Rather, it is the fate of the central core that captures our interest below.

Neutron stars are remarkable structures. As their progenitor stars were large and bright, so neutron stars are small, being perhaps a few tens of kilometers across, very

Fig. 7.2 The Crab Supernova remnant. This particular supernova event was observed by numerous scribes and court astronomers in the year 1054, and the nebula we see now is how it appeared some 6,500 years ago, at least according to its distance being 6,500 light years. At the center of the nebula is a pulsar composed of a small neutron star spinning on its axis some 30.2 times per second. (Image courtesy of NASA/HST)

faint, and extremely dense. Matter is so tightly packed within their deep interiors that the neutrons are literally squeezed to their Pauli Exclusion limit. It is the workings of this quantum mechanical phenomenon that keeps gravity at bay and stops neutron stars from collapsing straight into black holes – up to a point, that is.

Neutron stars are so very small and faint that they are difficult objects to directly observe with optical telescopes. The one notable exception to this rule, however, is the Geminga neutron star, which is situated some 200 parsecs from us (Fig. 7.3). It is more common to detect neutron stars with radio and X-ray telescopes, since they are known to be the powerhouses at the centers of pulsars – enigmatic objects that are nature's short-order timepieces (Fig. 7.4).

The existence of a pulsar, and by proxy a neutron star, is betrayed by the reception of rapid periodic pulses of electromagnetic emission. The radio signal is produced via synchrotron emission – the bugbear, recall, of the LHC tunnel designers – produced when electrons are accelerated to relativistic velocities along magnetic field lines. The magnetic field in question is actually locked to the neutron star's interior and is accordingly forced to rotate in step with the neutron star as it spins. In general, the magnetic field axis will be offset from the neutron star's spin axis, and accordingly the region from which the synchrotron radiation is emitted is swept around the sky like a lighthouse beam. Only if Earth chances to be located within the narrow emission beam will the pulsar be detected (Fig. 7.5).

The Geminga neutron star is known to be spinning on its rotation axis at an incredibly fast rate of once every 0.237 s. Other pulsars have been detected with spin rates ranging from a minuscule 0.0014 s to a stately 8.5 s. That neutron stars can spin so fast and not fly apart is a direct consequence of their very high compaction.

Fig. 7.3 XMM-Newton
Spacecraft image of
Geminga. The *arrow*
indicates the direction of
motion, and two X-ray tails
can be seen streaming behind
the object as it plows through
the surrounding interstellar
media at a speed of about
120 km/s. (Image courtesy
of ESA)

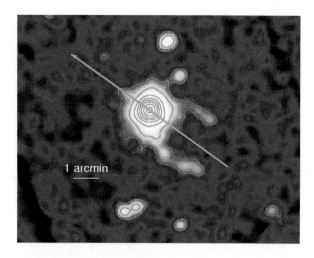

Fig. 7.4 Chandra X-ray
satellite image of the Vela
pulsar. The neutron star
located at the core of this
supernova remnant is
spinning once every 89 ms.
(Image courtesy of NASA)

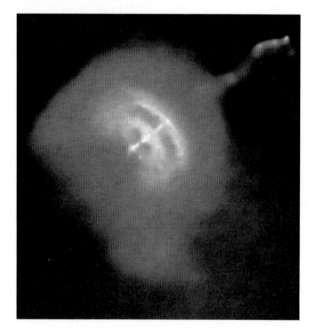

An isolated neutron star embedded within the core of a pulsar will slowly lose
rotational energy to its surrounding nebula (Fig. 7.2) and thereby gradually spin
down with increasing age. As time proceeds the neutron star will also radiate its
residual heat energy into space, becoming ever cooler. Since neutron stars do not
have to be hot in order to remain stable against collapse, however, this steady cooling

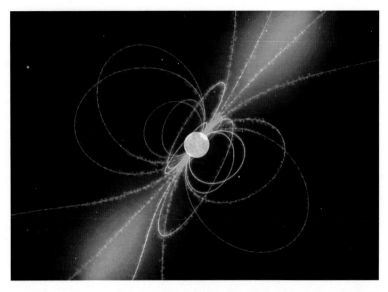

Fig. 7.5 A schematic diagram of the pulsar lighthouse model. In this illustration clouds of charged particles move along the magnetic field lines (*blue*) and create a narrow beam of gamma rays (*purple*) that are swept into our line of sight as the central neutron star rotates. (Image courtesy of NASA)

is of no great significance to their internal structure. Just like old university professors, isolated neutron stars never quite go away; they remain inflated and haunt the corridors of their youth.

Eventually, given eon piled upon eon, an isolated neutron star might collide with another star – possibly even another neutron star. It is only through these rare, distant in time, chance encounters that the demise of an isolated neutron star will be initiated. By accreting mass during a collision the neutron star won't necessarily be revived; rather it will be destroyed as gravity tears through the quantum mechanical limit, set by the Pauli Exclusion Principle operating upon the internal neutrons. With the exclusion pressure limit breached, gravity wins, and the neutron star will be crushed out of existence and become a black hole.

The mechanics of this final collapse are not well understood, and while the event is certainly terminal and tremendous amounts of energy are released, the full details are presently beyond the capabilities of present-day physics to fully explain. Although isolated neutron stars can be considered stable for exceptionally long periods of time, a neutron star that chances to have formed within a binary system might undergo a much more rapid demise.

In spite of the great upheaval associated with their formation, some neutron stars are born into binary companionship. Neutron stars are certainly known to exist within double star systems. This is perhaps a little surprising, since the only known mechanism for producing neutron stars is that of a supernova disruption, and under these circumstances it might be expected that the binary system should be ripped apart. This kicked-out-of-its-orbit scenario most likely explains the origin of high-speed objects such as Geminga (Fig. 7.3).

Even more remarkable, binary systems composed of two neutron stars are also observed; such binaries must be derived from remarkably lucky systems, since two supernova events have failed to destroy the pairing of the original massive stars. Under normal circumstances, however, and where all this stellar companionship becomes even more interesting, is that as the initially lower mass component of the binary star system evolves it can transfer matter onto its neighboring neutron star, thus driving up its mass.

This mass transfer scenario is believed to be responsible for the production of rapidly spinning, millisecond pulsars. Eventually, however, as matter continues to pile up on the surface of the neutron star, its stability limit will be overcome, and once the mass exceeds some 2 or 3 solar masses catastrophic collapse is inevitable.

With collapse, another explosion rocks the binary system. The neutron star will most likely become a black hole, and it may or may not remain bound to its mass-supplying companion. Remarkably, several X-ray binary systems are known in which a black hole companion appears to be in orbit around an accompanying mass-supplying, luminous star. The classic such system is Cygnus X-1, the first X-ray emitting source to be discovered in the constellation of Cygnus (Fig. 7.6). There is no clear consensus as to the exact mass of the unseen companion in Cygnus X-1, but some researchers place it as high as 8.5 times that of the Sun. With this mass the unseen companion cannot be a neutron star and must, therefore, be a black hole. The optically observed star within Cygnus X-1 has a mass that is about 25 times

Fig. 7.6 An artist's impression of Cygnus X-1. The binary system is composed of a massive *blue* supergiant star and a compact 8.5-solar-mass *black hole*. As gas streams away from the companion star it forms an accretion disk around the *black hole*. Before the matter finally spirals across the *black hole's* event horizon it is heated to very high temperatures and consequently emits copiously at X-ray wavelengths. (Image courtesy of ESA)

larger than that of the Sun, so it, too, will eventually undergo supernova disruption. Massive binary star systems lead very lively and eventful lives!

In addition to producing stellar mass black holes through accretion onto neutron stars within massive binary systems, it is also possible, but by no means proven, that some supernova events produce black holes directly. Astronomers have taken to calling these extreme collapse events hypernova, and they are thought to be associated with the production of so-called gamma-ray bursters – short duration, very high energy events that are detected in the extremely short wavelength, gamma ray region of the electromagnetic spectrum. Hypernova are believed to be associated with the collapse of massive 100-solar mass and larger stars (Fig. 7.7).

Fig. 7.7 The closest known hypernova candidate is Eta Carina – a luminous *blue* variable star that weighs in at an estimated 100–150 times the mass of our Sun. Located approximately 7,500 light years from us, when this star undergoes hypernova disruption it will appear, to terrestrial observers, as the brightest object in the sky after the Sun, full Moon, and planet Venus. (Image courtesy of NASA/HST)

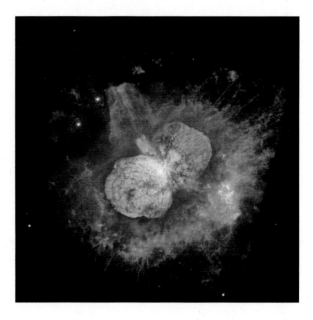

When all is said and done, astronomers presently believe that the end stage of a massive star's evolution is supernova disruption, leading to the formation of a neutron star or, in the more extreme cases, a black hole. A neutron star may also form in a binary system and thereafter accrete matter from its companion, a process that results in its mass steadily increasing with time. Eventually, in such binary systems, the mass of the neutron star is pushed over its stability limit, and it collapses into a black hole. All this being said, however, it is not presently clear where the upper mass limit of a neutron star sits; most researchers place the upper limit in the range of between 2 to 4 times the mass of our Sun. Although the upper limit to a neutron star's mass is not known to within a factor of two, the various theoretical models all predict relatively similar minimum radii of about 10 km. It is this latter radius limit that has some particularly interesting consequences, given the recent observations relating to an isolated neutron star with the less than romantic name of RXJ1856.5-3754.

The Strange Case of RXJ1856.5-3754 and Pulsar 3C58

Isolated neutron stars are rare objects, and less than a dozen such cosmic rarities have been identified. RXJ1856.5-3754 is the closest isolated neutron star to the Sun, and it was only discovered in 1996. At an estimated distance of 120 pc, RXJ1856.5-3754 is close enough to be imaged optically. Indeed, RXJ1856.5-3754 is moving through the interstellar medium so rapidly, speeding along at an estimated 200 km/s, that it has generated a bow shock (Fig. 7.8).

Fig. 7.8 The bow shock (*circled in the image*) generated by RXJ1856.5-3754 as it speeds through the interstellar medium at an estimated 190 km/s. (Image courtesy of ESO)

By combining optical and X-ray data it has been possible to estimate the surface temperature and size of RXJ1856.5-3754. The surface temperature turns out to be about 1 million degrees, which is not an unreasonable temperature for a relatively young neutron star (although we don't actually know when RXJ1856.5-3754 was formed); its estimated diameter, however, is an incredibly small 11 km.

The problem with this diameter is that it is half the size of the minimum set by the standard theoretical models. Somehow or other, RXJ1856.5-3754 has managed to cram an estimated 0.8 solar masses of material into an object having a radius of just 5.5 km, indicating a staggering bulk density of a 2.5×10^{18} kg/m^3 – equivalent to squashing the entire Earth into a sphere with a diameter of 170 m!

With its incredibly small size and extremely high density, RXJ1856.5-3754 is an enigma. Its observed characteristics do not match with those expected from the standard theory of neutron star structure, and accordingly it has been suggested that RXJ1856.5-3754 might be a strange star. The strangeness is not related to any specific peculiarity of RXJ1856.5-3754, however, but rather addresses the possibility that its interior might be composed of strange quark matter.

Almost immediately after the quark model was first elucidated by Gell-Man and others the suggestion was made that strange quark matter composed of roughly equal numbers of *u, d,* and *s* quarks might exist. Stable quark stars, it was argued, might come about following a quark–gluon plasma (QGP) phase transition within a massive neutron star's interior. It is not presently known, however, when such a transition might occur, or even if such a phase transition is possible. The picture may possibly become clearer once the LHC experiments to study the transition to a QGP state are completed. Figure 7.9 illustrates the key difference between the interior of the standard neutron star model, where the quarks are all confined to nucleons, and a strange quark star where the quarks are free and no longer confined. In essence a strange quark star can be thought of as a single massive atomic nucleus held together by the strong nuclear force.

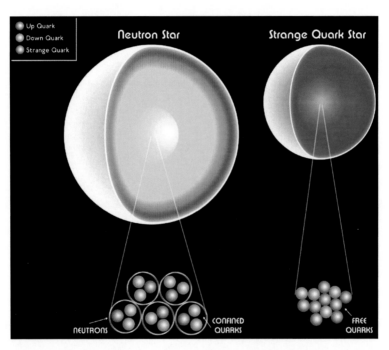

Fig. 7.9 In a neutron star the quarks are confined within the nucleons – a proton being the assembly {uud}, with the neutron being {udd}. After the QGP transition the quarks are no longer confined to nucleons, and a strange quark star might be produced. (Image courtesy of NASA)

The surfaces of such objects would be very strange to encounter, with the density jumping from the near vacuum of interstellar space to some 10^{17} kg/m^3 in the span of just 1 fermi (10^{-15} m) – talk about hitting a brick wall. The properties of quark stars are such that they are, for a specified mass, much smaller in size than a neutron star, and it is for this reason that some theoreticians have suggested that RXJ1856.5-3754 might be our first glimpse at a new kind of stable strange matter.

The explosive event that produced the supernova remnant known as 3C58 (the 58th object in the third Cambridge radio telescope survey) was first observed from Earth on the August 6, 1181. Ancient records kept by Chinese court astronomers reveal that upon this particular date a "guest star appearing in k'uei invading the Ch'uan-she-hsing" was observed. In the Western tradition this places the guest star (or supernova) in the constellation of Cassiopeia. The Chinese chronicles further record that the new star was visible for over 185 days.

The estimated distance to the 1181 supernova remnant is 3.5 kpc, and observations indicate that the nebulosity has expanded to cover a region having dimensions of about 11-pc by 7-pc (Fig. 7.10). The nebula superficially resembles the Crab Nebula (associated with the supernova of 1054 – see Fig. 7.2), and like the Crab Nebula, 3C58 was derived from the collapse of a massive star's iron core (making it a Type II supernova). Within the core of 3C58 is a pulsar, spinning at a relatively sedate 15 times per second. Figure 7.11 shows the X-ray brightness of the 3C58 pulsar, and the short-duration pulses, produced as the high-energy synchrotron beam sweeps across our line of sight, are clearly identifiable.

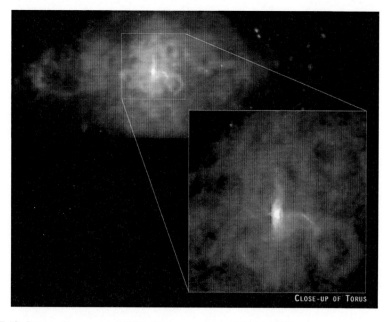

Fig. 7.10 Image of neutron star 3C58. (Image courtesy of NASA/HST)

In early 2002 astronomers were, for the first time, able to use the X-ray brightness data to estimate the surface temperature of the pulsar's underlying neutron star – and they were immediately surprised by their findings. The surface temperature was a little over 1 million degrees, but this, importantly, was much cooler than the standard neutron star model predicts. Although it is to be expected that neutron stars cool off with time, it was also believed that they should do so very slowly. Remarkably,

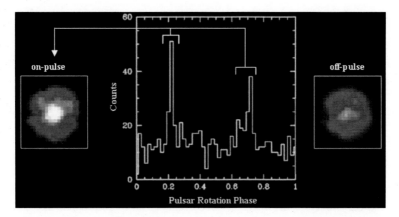

Fig. 7.11 The rotation of 3C58 as revealed by observations collected with the Chandra X-ray telescope. The on phases are indicated by the signal spikes at phases corresponding to 0.2 and 0.7. The X-ray appearance of 3C58 during the on and off phases are shown in the smaller images to the *left* and *right* in the diagram. (Image courtesy of NASA)

the 3C58 neutron star had, after just 820 years, cooled to a temperature that would otherwise be expected after some 20,000 years. Something was clearly amiss.

Neutron stars cool off through the emission of neutrinos, produced when neutrons and subatomic particles interact within their deep interiors. The cooling rate depends critically upon the interior density and composition, and the apparent rapid cooling of 3C58 suggests that it might not be a typical neutron star, but may in fact be a smaller, much denser strange quark star, just like RXJ1856.5-3754.

Looking a little further afield than 3C58, astronomer Kwong-Sang Cheng (University of Honk Kong, China) and colleagues have recently suggested that the naked-eye supernova that occurred in the Large Magellanic Cloud in 1987 (SN1987A) might also have produced a quark star. Their argument is based upon the neutrino flux data recorded at the Kamiokande II observatory in Japan and at the Irvine-Michigan-Brookhaven neutrino detector in the United States. What they claim is that the combined data set indicates a significant delay of several seconds between neutrino bursts being detected at the two observatories. This delay, however, is not related to the geographical distance between the detectors but rather indicates, Cheng and colleagues argue, that SN1987A gave off two distinct neutrino blasts. It is the second neutrino burst that Cheng and co-workers suggest is the signature of the neutron star collapsing into a strange quark star.

Well, so much for the background astronomy. If strange quark stars really do exist, however, then it is just possible that smaller nuggets of strange quark matter, generally referred to as stranglets, might also exist. Stranglets are purely hypothetical objects; none have ever been detected, but the suggestion that they might be generated at the RHIC and the LHC has resulted in several safety review commissions being formed. The danger, it has been argued, relates to the possibility that if a stranglet forms then it might cause the catalytic conversion of all surrounding matter, i.e., Earth and us, into strange quark matter nuggets. Clearly, even the generation of one stranglet could be catastrophic.

Under normal circumstances any particle containing a strange (*s*) quark decays rapidly into lighter particles containing only up (*u*) and down (*d*) quarks. However, it was suggested initially by Arnold Bodmer (Professor Emeritus, University of Illinois, in Chicago) in the early 1970s and later by Edward Witten (Institute for Advanced Studies, Princeton) in the mid-1980s that large numbers of quarks in certain states might produce stable matter. Indeed, under what has become known as the strange matter hypothesis (SMH), it is possible that all nuclei will eventually decay into stranglets because stranglets have a lower ground state energy.

The conversion of nuclei into stranglets, however, will be (if the SMH is true) a very, very slow process, since many *s* quarks will need to be generated simultaneously, which is an extremely unlikely occurrence even on timescales as long as the age of the universe. In collider experiments such as those conducted at Brookhaven's RHIC where numerous large nuclei interactions take place at relativistic velocities, strange and antistrange quarks could conceivably be produced, possibly resulting in the brief formation of a stranglet. The experimental signature of such an oddity would be a particle with a very distinctive and very obvious high mass-to-charge ratio, but no such particle has ever been detected by RHIC researchers. The operating characteristics of the LHC, it turns out, are even less likely to produce stranglets than the RHIC experiment, and while the LHC will certainly produce strange quarks it will produce them with such high energies that the probability of them accumulating to produce a stranglet is next to zero – if not zero itself.

Not only is the generation of stranglets highly unlikely at the LHC, it can also be argued that even if one or more such objects did form, then a disaster scenario would only play out if the stranglets had a net negative charge. This charge requirement is in contrast to expectations, given that theory suggests that stranglets should naturally have a positive charge.

Indeed, the standard scenario envisioned is that a stranglet will rapidly surround itself with a cloud of electrons (adopting the characteristics of a bizarre atom). A positively charged stranglet would fail to interact with ordinary matter, since it would experience the Coulomb repulsive force – the same force that stops ordinary nuclei from spontaneously collapsing at low energies.

There are many "ifs" involved in the chain of reasoning leading to the stranglet disaster scenario, and our present understanding of physics makes each one of them exceptionally unlikely. The destruction of Earth through the production of stranglets at the LHC is one of the least likely ways that our existence will come to an end. Indeed, our continued safety against suffering a stranglet death is writ large across the sky in the form of our companion Moon (Fig. 7.12). Cosmic rays (to be discussed more fully later) have been striking the Moon's rocky surface for the past 4.5 billion years, and many of these cosmic rays will have had impact energies far in excess of those that will be realized by the LHC. The fact that the Moon hasn't collapsed into a strange quark globe argues strongly, therefore, for the fact that the LHC is perfectly safe with respect to the initiation a world-altering stranglet event. Indeed, in March 2009 Ke Han (Yale University) and collaborators reported

Fig. 7.12 Apollo 16 image of the Moon's far side. If strangelets were truly able to cause matter conversion, then because its surface is directly exposed to space, the Moon should have collapsed into a strange quark sphere billions of years ago as a result of cosmic ray impacts. (Image courtesy of NASA)

on an experiment to look for strangelets embedded within lunar soil. Using 15 g of material collected during the Apollo Lunar Missions the soil grains were fired past a strong magnet and their paths followed. If a specific grain contained a stranglet then its path would curve less than those followed by ordinary (stranglet-free) grains. No stranglet hosting grains were detected.

Small, Dark, and Many Dimensioned

If Earth and humanity are not to be destroyed by the catastrophic bang of stranglet conversion, what are the chances that it might end with a faint whimper as it slips past the event horizon of a black hole? Although the Milky Way Galaxy almost certainly contains numerous stellar mass black holes, as well as at least one super-massive black hole at its very core (Fig. 7.13, and recall Fig. 7.6), the likelihood of Earth actually encountering such objects is negligibly small on timescales of the age of the Solar System (a time of order 4.56 billion years) and probably much longer.

However, what might be the consequences for Earth if a small black hole were to be produced in the LHC experiment? This disaster scenario plays from the inside outward. Assuming low mass black holes (the properties of which we shall look at shortly) are produced within the LHC, then they will have near-zero relative velocity with respect to Earth's surface and accordingly they should sink under gravity towards our planet's core. As the small black holes accumulate and merge at Earth's center, so the argument runs, a larger and much more deadly black hole might be

Fig. 7.13 Sagittarius A* is
situated at the core of our
Milky Way Galaxy, some
26,000 light years from the
Sun. The radio bright source
known as Sgr A* actually
surrounds a 3.7-million solar
mass black hole. This
Chandra X-ray telescope
image of the galactic center
reveals numerous, multiple,
light-year-long looping
clouds of hot gas surrounding
the central black hole. (Image
courtesy of NASA)

produced. With continued growth the central black hole will begin to consume its
surroundings, pulling more and more of Earth's interior into its hungry maw. The
accretion process will eventually become exponential and soon thereafter, in one
final rush, Earth will blink out of existence. Clearly consumption by black hole is
not an especially pleasant way for the world, and all of humanity, to end, but is the
black hole production and accumulation scenario a realistic one?

Under slightly different circumstances than those that apply here, the author,
along with other researchers, have described methods by which black holes might be
employed by our distant descendants to generate energy and to aid in the terraforming of new worlds, but these scenarios play out under conditions set far beyond
those that can be achieved today. Black holes may well have a utilitarian function
in the distant future, but in the here and now the close proximity of any such object
can only be viewed as a threat. All this being said, the possibility of Earth being
destroyed by the generation of multiple miniature black holes at the LHC is, to say
the least, highly remote.

Black holes are regions of our universe that have become cut off from direct
communication. What information passes through the event horizon of a black hole
is lost to us forever. The intense gravitational field that defines and characterizes
a black hole warps the surrounding space so violently that nothing, not even light,
can escape from within its deep interior – the boundary being a spherical surface
described by the Schwarzschild radius.

Introduced by German physicist Karl Schwarzschild in 1916 as part of his solution to Einstein's equations of general relativity, the Schwarzschild radius r_s of a
non-rotating black hole of mass M is given by the relationship $r_s = 2\,G\,M/c^2$, where

G is the universal gravitational constant and c is the speed of light. Any object with a radius $r < r_s$ must undergo total gravitational collapse to form a black hole. The interior of a black hole, that is, the region interior to radius r_s, is a domain profoundly cut off from our direct examination and one in which we know that both new and extreme physics must eventually prevail.

For the Sun, with a mass of some 2×10^{30} kg, the Schwarzschild radius is about 3 km (Fig. 7.14), while for Earth, which weighs in at about 6×10^{24} kg, it is about 9 mm. Imagine the sheer compression required to cajole the entire Earth into a sphere smaller than the size of a golf ball. Remarkably, if it wasn't for the repulsive Coulomb force experienced between atoms then gravity would do just this. Indeed, if nothing prevented it from doing so, Earth would collapse under it own gravity to form a black hole in about half an hour.

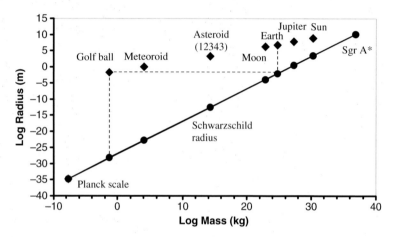

Fig. 7.14 Size versus mass diagram for various astronomical objects (and a golf ball). The *solid diagonal line* indicates the Schwarzschild radius beyond which a given mass must collapse to produce a black hole. The *diamonds* indicate the present size and mass of the Sun, the Moon, Asteroid 12343 martinbeech (the author's favorite minor planet), and a 1-m diameter stone meteoroid. The *vertical dashed line* from the Earth symbol indicates the amount of compression required to turn Earth into a black hole. A black hole with the mass of Earth will have a Schwarzschild radius similar to that of a golf ball (*horizontal dashed line*)

Under astronomical circumstances the presence of a black hole can be deduced by the heating of matter as it swirls around the event horizon (recall Fig. 7.6) – the gas glowing violently before it disappears for eternity from our universe. Ultimately, as we saw in Chapter 6, most of the matter within our Local Group of galaxies will end up within the confines of black holes. In a gravitational sense our future is deeply compressing. All the above being said, the LHC might, at best, produce only very small and very low mass black holes. More remarkably, perhaps, if the LHC does produce any such objects it will immediately tell us that the universe must be composed of more than the familiar 3-spatial plus 1-time dimensional spacetime we experience in our everyday lives.

Under the paradigm of the Standard Model the smallest mass black hole that might conceivably be generated will have a mass of about 2×10^{-8} kg (the Planck mass) and an event horizon corresponding to the Planck length of about 1.6×10^{-35} m. It is on the Planck scale, recall, that the ultimate quantum nature of gravity must come into play. The energy associated with a Planck mass black hole will be the Planck energy corresponding to $E_P \approx 2 \times 10^9$ J $\approx 10^{16}$ TeV, and accordingly with a maximum collisional energy of 14 TeV the LHC operates well below the threshold for producing even the lowest possible mass black hole.

Where this standard argument changes, however, is when we consider the possibility of highly compacted extra dimensions. In this situation standard particle physics is confined to work within the familiar 4-D (that is 3 spatial + 1 time dimensional) domain. This 4-D world is referred to as "the brane," which is short for membrane. In contrast to the particle physics domain, gravity is assumed to act over many more extremely small and compact spatial dimensions – dimensions that we simply can't see in our 4-D brane. The extra dimensions over which gravity is allowed to operate are rather unceremoniously called "the bulk" (Fig. 7.15).

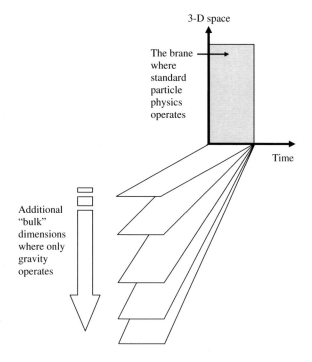

Fig. 7.15 A schematic, minds-eye, view of the idea behind extra dimensions. The vertical (*shaded*) plane is "the brane," and this is where you and I live and experience the world, and it is also where experimental physics can be performed. The additional planes represent the "bulk" dimensions over which gravity operates. Note that the compaction and curvature of the bulk dimensions are not illustrated in this diagram

This brane + bulk model was introduced in an attempt to explain why gravity is such a weak force compared to all the atomic forces. The exact details need not concern us here, but according to some extra dimensional theories the energy scale upon which quantum gravity effects will become noticeable are as low as 1 Tev, and it is possible, therefore, that the LHC might just produce low mass black holes.

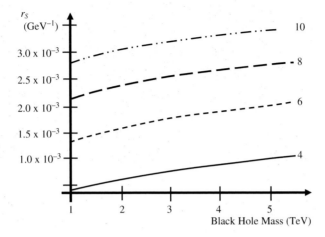

Fig. 7.16 Schwarzschild radius versus the mass of a non-rotating black hole for a range of additional spatial dimensions. The various *curves* are labeled according to the number of dimensions, with the *solid line* (labeled 4) showing the standard 3 spatial + 1 time dimensional spacetime situation

Within such extra dimensional theories the Schwarzschild radius for a black hole increases both with its mass as well as with the number of extra dimensions (Fig. 7.16). Model calculations indicate that if there are, for example, four additional bulk dimensions, then the Schwarzschild radius will increase by a factor of about 5.

If, and it is still a very big if, the LHC manages to produces low mass black holes, what might they do? It has been suggested that if produced in large enough numbers the low mass black holes leaking out of the LHC might go on a runaway feeding frenzy, consuming Earth atom by atom at first, but eating ever more larger pieces as their mass and Schwarzschild radii increase. The feeding frenzy would continue until the entire Earth had been devoured.

This consumption scenario, however, for all of its catastrophic threat, is almost certainly wrong – and wrong according to several good lines of argument. First, a detailed study published towards the close of 2008 by Steven Giddings (University of California, at Santa Barbara) and Michelangelo Mangano (CERN) found that microscopic black holes simply can't attain a size sufficient to realize macroscopic proportions under any realistic accretion scenario. This result comes about since, for a particle to be accreted, it will have to encounter the black hole with an impact parameter smaller than the Schwarzschild radius. The rarity of such fine-tuned events occurring results in a very slow growth rate. Indeed, the growth rate is so slow that it takes longer than the present age of the Solar System for a Planck mass black hole to grow to a size where it might be dangerous to the continued existence of Earth.

Later on in this chapter we will discuss the properties of cosmic rays, rapidly moving charged particles and their interaction with objects within the galaxy at large. Collisions, however, between cosmic rays and the matter within stars constitute nature's equivalent of a collider experiment, and cosmic rays, it turns out, can pack a punch even more energetic than those that will be achieved at the LHC. Picking up on this observation Michael Peskin (Stanford University) noted in a review article published in the journal *Physics* for August 2008 that the continued existence of "super-dense stars [i.e., neutron stars] act as the proverbial canaries in the coal mine."

His argument runs along these lines. If a very high energy cosmic ray collision produces a small black hole at the surface of a neutron star, then a very high accretion rate should result, since the matter in a neutron star is compacted to an extremely high density. This naturally high accretion effect will result in the rapid consumption of the neutron star. Since, however, we know that neutron stars and pulsars exist, it can be concluded that either Planck mass black holes are not produced in cosmic ray collisions, or, if they are produced by cosmic ray collisions, then they cannot accrete efficiently and are accordingly harmless. The monotonic radio pips of galactic pulsars, therefore, sing-out in one vast cosmic chorus the continued safety of Earth against miniature black hole consumption.

The accretion scenario studied by Giddings and Mangano is just one line of argument, and it was deliberately based upon the unlikely (but theoretically imaginable) scenario that any mini black holes produced by the LHC would be stable. That microscopic black holes should be unstable, and in fact evaporate away on extremely short timescales of order 10^{-24} s, is based upon the idea of Hawking radiation. This concept, which is almost universally accepted although not as yet experimentally proven, was developed by Stephen Hawking (Cambridge University, England) in a series of now classical research papers published during the 1970s. In his research papers Hawking demonstrated that black holes should exhibit a number of remarkable features.

First, black holes are not black, and secondly they are not (in spite of our earlier comments) eternal. That black holes do emit radiation into space comes about because of the strong gravitational effects that operate near the event horizon and because of quantum vacuum fluctuations. We encountered the idea of the Heisenberg uncertainly principle in Chapter 1, and by recasting it in a form that relates energy and time we can unravel the origin of Hawking radiation. The idea of energy conservation, as discussed in Chapter 1, holds supreme within classical physics: energy is neither created nor destroyed within a closed system – it can only change from one form to another.

In the quantum world, however, energy can be instantaneously created, or borrowed, but only for a short amount of time. The uncertainty relationship between the amount of borrowed energy ΔE and the time before it must be paid back Δt is $\Delta E \Delta t > \hbar$. Here we see that the greater the amount of energy borrowed so the shorter must the time interval be before it is paid back. Importantly, what this relationship tells us about the vacuum of space is that it is never actually devoid of particles, but it is rather a seething broth of virtual particle-antiparticle pairs jumping fleetingly into and then out of existence. In essence, by borrowing energy ΔE a particle-antipaticle pair can be created by a quantum fluctuation. The borrowed energy is later returned, however, when the particle and antiparticle annihilate each other after a time interval $\Delta t \sim \hbar/\Delta E$.

Where all this fleeting particle formation becomes interesting, however, is if the quantum fluctuation occurs near to the event horizon of a black hole. In this case, it is possible that the particle-antiparticle pairing can be broken, with one particle falling through the Schwarzschild radius (becoming therefore lost to our universe and trapped within the black hole) and the other particle, now no longer having its antiparticle with which to be annihilated, is ejected from the event horizon region into space. Importantly, as we shall see in a moment, the particle that enters the black hole carries negative energy, while the particle that is radiated away from the event horizon has positive energy. A minds-eye picture of the Hawking radiation process is presented in Fig. 7.17.

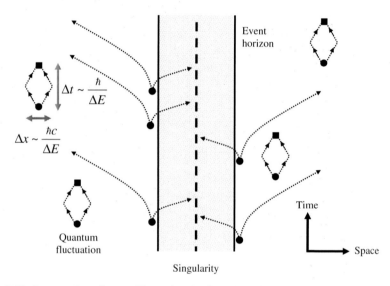

Fig. 7.17 A space-time diagram illustrating the formation of virtual particle-antiparticle pairs through quantum fluctuations. In this diagram time is imagined to increase in the vertical direction, while 3-D space is reduced to just horizontal variations across the page. The *filled circles* indicate the time and location at which a particle pair is created in a vacuum fluctuation, while the *filled squares* represent the time and location of their termination. The *dashed heavy line* represents the central singularity of the black hole and the *solid vertical lines* represent the event horizon. The *arrowed* and *dotted lines* illustrate the trajectories of the particle pairs that form close to the event horizon

Not only should black holes emit radiation into space, they must also have a related temperature, Hawking showed. The temperature of a black hole varies inversely with its mass; the more massive a black hole is the lower must its associated temperature be. The effect of extra dimensions adds a complication to the temperature determination, but in general the greater the number of dimensions, the greater the Hawking temperature should be for a given black hole mass.

As a direct consequence of the particle-antiparticle separations occurring at the event horizon, conservation of energy requires that the particle that enters the black hole must have negative energy. The result of this is that the mass of the black hole is reduced. We can think of this in terms of Einstein's mass-energy equivalency formula: $E = mc^2$, where E is the energy, m is the mass, and c is the speed of light. Since the c^2 term must always be positive, a negative energy corresponds to a mass deficit. Through the Hawking radiation process, therefore, black holes must ultimately evaporate away their mass.

The evaporation time T_{evap} for a black hole varies as the cube of its mass M. The bigger the mass of the black hole, the longer it takes to evaporate away. If the mass of the black hole is given in kilograms, then the detailed theory gives $T_{\text{evap}} = 8.407 \times 10^{-17} M^3$ s. For a supermassive black hole, such as that located at the center of our galaxy (Fig. 7.13), the evaporation time is gargantuan. Indeed, Sgr A∗, which has a mass of 3.7 million solar masses (equivalent to 7.4×10^{36} kg), has a decay time, provided it doesn't continue to accrete additional stellar material, of some 3.4×10^{94} s, or a colossal 10^{87} years. In contrast, a 230,000-kg black hole (having a Schwarzschild radius of 3.4×10^{-22} m) will evaporate away in about a second. A Planck mass ($M_{\text{PL}} = 2 \times 10^{-8}$ kg) black hole will evaporate within 10^{-39} s.

The safety of Earth against consumption from low mass black holes produced by the LHC is assured on two counts. First, the accretion rate of any such black holes will be extremely small, and second their evaporation times against Hawking radiation are so short that they won't survive long enough to accrete even a handful of atoms. Given these circumstances the actual production of low mass black holes with the LHC is to be hoped for, since their detection will provide fundamental information and indeed confirm the existence of extra dimensions beyond those of our familiar 4-D spacetime. What an incredible thought that is.

In the multi-dimensional model of black hole formation, the Hawking radiation will emit Standard Model particles into the brane and gravitons into the bulk, and the tricky problem for the theoretician is to describe exactly how these two processes will play out. The problem, however, has recently been tackled by an international team of physicists headed up by De-Chang Dai (Case Western Reserve University, Cleveland). The team has developed a computer program, called *BlackMax*, to simulate the experimental signatures that might be seen at the LHC if microscopic black holes are produced. The computer model predicts that one of the key observational characteristics of low-mass black hole production will be their rapid decay into a very large mix of particles.

Indeed, in graphic terms, Greg Landsberg (Brown University in Providence, RI) has commented that the detectors at CERN will light up like a Christmas tree following the decay of a miniature black hole. Not only this, the particles produced during the decay should be emitted in an isotropic manner. Figure 7.18 shows one simulation for the possible decay of a low mass black hole in the ATLAS experiment. Under some decay modes a miniature black hole might produce many gravitons that will be radiated into the bulk (rather than entirely into the brane), producing an asymmetry in the direction of the emitted particles and an apparent energy deficit. Remarkably, by recording and characterizing the decay modes it might be possible to determine just how many additional dimensions there are in the universe.

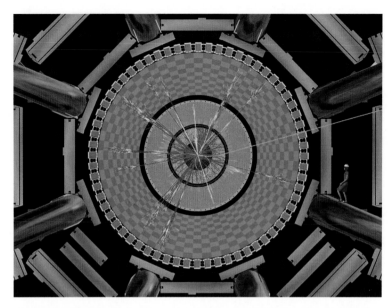

Fig. 7.18 Simulation of the decay signature that might result from the formation of a mini-black hole in the ATLAS detector. The view is looking down the beam pipe. (Image courtesy of CERN)

This Magnet Has Only One Pole!

Just as shoes are always brought in pairs, one for the left foot and one for the right, so magnets always have a pair of poles – a north pole and a south pole. Take any standard bar magnet and cut it into as many pieces as you like; no matter how small you might make the fragments, they will always have a north pole and a south pole. There is no limit beyond which the cuts suddenly start producing only north or only south magnetic polarities. Magnetic poles always come in pairs, just like shoes. Or do they?

Well, the answer to this is yes, they are always found in pairs, and no experiment has ever produced a clear, unambiguous detection of a magnetic monopole – an object that has just one magnetic polarity. Although this is the current experimental status, it is philosophically a rather poor state of affairs, since we cannot prove the non-existence of an entity purely on the basis of saying "we've never seen one."

British physicist Paul Dirac (Fig. 7.19) first discussed the idea of magnetic monopoles in 1931, and they have been a much investigated phenomena ever since. The suggestion that particles having just a north (or a south) magnetic polarity might exist pays homage to the notion of symmetry within the workings of nature; if particles having just a positive or a negative electrical charge can exist, then why not particles with just a north or a south magnetic pole? In his original paper Dirac used the existence of magnetic monopoles to explain why electrical charge must be quantized, but in more recent times their existence has been linked to the possibility that the electromagnetic, weak nuclear and the strong nuclear force might converge into

Fig. 7.19 Paul A. M. Dirac (1902–1984). One of the towering intellects of mid-twentieth-century physics

one unified force at very high energies. Such grand unified theories (GUTs) tend to predict the existence of large numbers of monopoles, their origin being linked to the earliest moments of the Big Bang when our universe first came into existence.

Many experimental searches for magnetic monopole have been made over the years, and one such past experiment was conducted at CERN with the SPS (super proton synchrotron) in the early 1980s.

In the CERN experiment high energy protons were fired at targets containing ferromagnetic tungsten powder. Any magnetic monopoles produced in the collisions, it was argued, should become trapped in the target powder. Placing a solenoid capable of generating a very strong magnetic field in front of the irradiated targets, it was then reasoned, should enable the extraction and ultimate detection of a magnetic monopole. None was found. More recently researchers using the proton-antiproton Tevetron collider at Fermilab have looked for magnetic monopoles that might have become trapped within the detector framework – ghosts, as it were, in the machine, but again no definitive signatures betraying the presence of monopoles were discovered.

The search for magnetic monopoles took an important step forward in September 2009 when two independently working research teams reported the detection of "discrete packets" of magnetic charge (regions with characteristics akin to that of monopoles) in special, super-cooled crystals known as spin ice. While not magnetic monopoles in the exact sense proposed by Dirac these new experiments have clearly demonstrated that isolated magnetically charged regions can exist, and further research, published in October 2009 by Stephen Bramwell (London Center for Nanotechnology) and co-workers, has shown that the magnetically charged "quasi-particle" regions in spin ice can be made to flow under an applied magnetic field – just like electrons flow when placed in an electric field. This new phenomena, dubbed magnetricity, may ultimately play a central role in the manufacture of magnetic memory storage devices and electronic computing.

Crepuscular, fleeting, elusive, and yet fundamental, the right experiment to demonstrate the existence of Dirac's magnetic monopoles hasn't, as yet, been performed. Be all this as it may, it is still always possible that the LHC might be the right machine with the power to do the job.

As with the production of strangelets and low mass black holes there are also possible problems associated with the generation of copious numbers of magnetic monopoles – again, the fear is that the LHC might trigger a catastrophe. At issue is the potential catalytic destruction of protons.

In the Standard Model the proton is absolutely stable and, even if left to its own devices, will never decay into lighter particles. Although proton decay is possible in some non-standard models, there is presently no evidence to indicate that such decays actually occur. The best experimental constraint on the half-lifetime, the time a given quantity of protons is reduced by a factor of two, is provided by the Super-Kamiokande Cerenkov detector in Japan (Fig. 7.20). And the estimated half lifetime is greater than a monstrous 10^{35} years.

Various estimates for the masses of magnetic monopoles exist, but they are generally expected to weigh in at a colossal 10^{12} TeV – far too heavy to be produced by the LHC. Nonetheless, given all of the theoretical uncertainties about their actual properties the possibility that the LHC might produce monopoles is worth considering. Indeed, the MoEDAL (Monopole and Exotics Detector at the LHC) experiment will look for monopoles and highly ionizing stable massive particles produced in association with the LHCb experiment (Chapter 4). The passive MoEDAL detector is composed of numerous nuclear track detector sheets that will be placed close to the vertex locator of the LHCb. Due to be completed in the winter of 2010 the MoEDAL experiment will take year-long exposures of particle production activity at the LHCb with the hope of capturing distinctive magnetic monopole tracks.

At issue with respect to the proposed risk of magnetic monopole (MM) production is their possible interaction with protons (P). A number of outcomes of such interactions are possible; for example the encounter might run according to the sequence: $MM + P \Rightarrow MM + e^+ + \pi^0$ (i.e., the proton decays into a positron and a neutral pion). Importantly in this interaction, while the proton is broken down into smaller particles, the magnetic monopole emerges unscathed, and it might therefore cause a second proton decay, and a third, and a fourth, and so on, resulting in the explosive release of energy and the catastrophic collapse of surrounding matter.

To see that such a scenario of destruction is highly unlikely, we need to look beyond the confines of Earth and investigate the properties of cosmic rays.

Fig. 7.20 Technicians inspect a row of photomultiplier tubes in the partially water-filled cavity of the Super-Kamiokande detector. The decay of a proton will produce a positron with a velocity that exceeds that of light in water, and under these conditions it will emit Cerenkov radiation. The photomultiplier tubes will detect the characteristic Cerenkov light, and thereby register the fact that a proton has decayed somewhere within the reservoir of water. (Image courtesy of Kamioka Observatory)

These Rays Are Truly Cosmic

The detonation of a supernova, as we noted earlier, heralds the end to the brilliant but short-lived life of a massive star. The iron cores of such doomed stars become neutron stars, black holes, or possibly quark stars, while their outer envelopes are blasted into the surrounding interstellar medium. A supernova event is tumultuous, chaotic, and terminal. Remarkably, however, astronomers now understand that supernovae are responsible for the birth of new stars. Even our Sun and Solar System owe their origin to the final death throes of a once-glorious stellar behemoth. Indeed, by plowing up the gas and dust in the interstellar medium, supernova can trigger star formation through the compression of otherwise cold and diffuse gas clouds.

That the origins of our own Solar System were the result of a supernova explosion is directly betrayed by the presence of the element ^{60}Ni in iron meteorites (Fig. 7.21). This particular isotope is derived from the radioactive decay of ^{60}Fe

Fig. 7.21 The dramatic display of Gibeon iron meteorite fragments that line Post Street Mall in Windhoek, Namibia. These meteorites contain the element nickel-60 that can only have been produced through the decay of iron-60. The iron-60, in turn, could only have been generated in a supernova explosion

through the conversion of two neutrons into protons via a series of inverse beta decays: $^{60}Fe \Rightarrow e^- + {}^{60}C \Rightarrow e^- + {}^{60}Ni$. Two antineutrinos are also emitted during this decay process. The time for half of a given sample of ^{60}Fe to decay into ^{60}Ni is a cosmically short interval of just 1.5 million years, indicating that after just a few tens of millions of years any ^{60}Fe originally present will have decayed into its daughter product of ^{60}Ni.

The fact that ^{60}Fe has such a short half-life and that ^{60}Ni is found within iron meteorites reveals that the ^{60}Fe must have been present and reasonably well mixed within the solar nebula from the very earliest of moments – even before the planets had formed. Further, since the only known source for ^{60}Fe is a supernova explosion, the burst of star formation leading to the formation of our Sun and Solar System must have been triggered by a supernova detonation. Not only, therefore, are we composed of atoms that were generated in myriad ancient supernova, but the life-nurturing Sun and the very planet upon which we live owe their beginnings to the explosive death of a massive star. We are truly the descendents of primordial chaos.

Not only do supernova blast waves stir up and mix the interstellar medium, they also launch particles into their surroundings at highly relativistic speeds. These cosmic bullets are mostly protons, but some are helium nuclei and others are free electrons. The net charge, either positive or negative, associated with these rapidly moving cosmic rays dictates that they must generally follow the magnetic field lines that twist and weave their way through the disk of our galaxy. They pervade the spiral arms and speed around the galactic disk, and some collide with Earth. Indeed, the cosmic ray flux links Earth to the greater galaxy and taps out a ghostly Morse code message of destruction – an obituary composed of hyper-energetic particulate dots and dashes speeding away from the violence of distant supernovae explosions.

When a cosmic ray proton chances to encounter Earth, it will soon interact with an atmospheric atom, producing a great cascade of millions of elementary particles (Fig. 7.22). Every such cosmic ray shower is just like a collider experiment being performed, for free, within Earth's upper atmosphere. Cosmic rays arrive at Earth with a whole range of energies, with the lowest energy impacts being much more common than higher energy ones. The measured flux of cosmic rays at Earth is illustrated in Fig. 7.23. The more common, low-energy cosmic rays are derived from the Sun, and typically there are one or two impacts from these cosmic rays per square meter per second at the top of Earth's atmosphere.

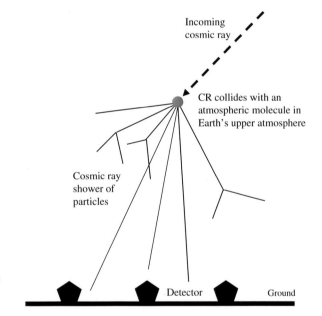

Fig. 7.22 A schematic cosmic ray shower. The collision between a cosmic ray and an atmospheric molecule produces a vast spray of particles that can be studied with an array of ground-based detectors

The galactic supernovae-produced cosmic rays are much more energetic than their solar counterparts, but they are much less common, producing a few hits per square meter per year at the top of Earth's atmosphere. The highest energy cosmic rays are believed to be derived from extragalactic sources (such as hypernovae in distant galaxies), and these arrive at Earth with a flux equivalent to several hits per square kilometer per year.

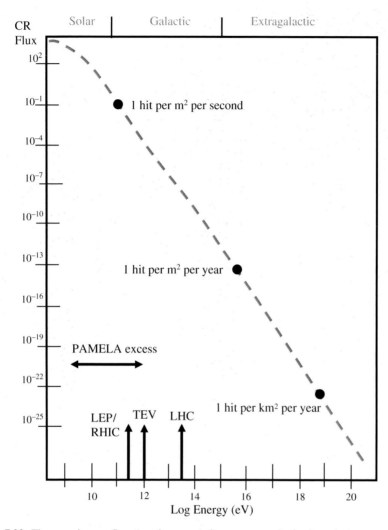

Fig. 7.23 The cosmic ray flux (number per unit area per unit time) against energy. The approximate source region(s) for the cosmic rays are indicated at the *top of the diagram*. The highest-energy cosmic rays are derived from outside of our galaxy. The approximate maximum collision energies for terrestrial collider experiments are indicated by *upward arrows*. The *horizontal arrow* shows the energy region over which the PAMELA spacecraft results indicate an excess of positrons reputedly due to dark matter annihilation events

The range of energies indicated in Fig. 7.23 reveals an important point about cosmic rays. Most of the galactic and all of the extragalactic cosmic rays will encounter Earth's atmosphere with energies far greater than the collisional energies that will be realized by the LHC. With respect to the safety concerns that the LHC might trigger the destruction of Earth, the highly energetic cosmic rays tell us that our worries are without foundation.

The observational data indicate that something like 5×10^{-10} cosmic rays with energies greater than 10^{17} eV strike each square meter of Earth's upper atmosphere every second. Given a total surface area of 5×10^{14} m^2 and an age of 4.56 billion years, something like 3.5×10^{22} cosmic rays with energies greater than 10^{17} eV have hit Earth's atmosphere since it formed. Because cosmic rays hit atmospheric molecules that are essentially stationary, a 10^{17} eV cosmic ray impact packs about the same amount of energy as a full power head-on, proton–proton collision within the LHC. Nature, therefore, has already performed, in Earth's upper atmosphere, a great multitude of equivalent LHC experiments.

When the LHC is operating at full energy it will produce about a billion proton–proton collisions per second. If we assume that the collider will run for about 10 million second per year (i.e., for about a third of a year) for each year of its anticipated 10-year lifetime, then the total number of collisions generated within the LHC will amount to about 10^{17} events. Through cosmic ray encounters, therefore, nature has already conducted some 350,000 LHC experimental runs since Earth formed – and, importantly for us, Earth still exists.

Indeed, there is absolutely no indication that any one of the 3.5×10^{22} cosmic ray events with energies greater than 10^{17} eV has ever produced any exotic physical phenomena that has had a detrimental effect on Earth. These numbers can be further bolstered if we consider the planet Jupiter. This giant planet has a surface area 128 times larger than that of Earth, and accordingly over the age of the Solar System some 45 million cosmic ray equivalent LHC experiments have been conducted within its upper atmosphere – and it still exists. The Sun has an even larger surface area again, being a factor of 12,000 times larger than Earth's, and accordingly some 4 billion cosmic ray equivalent LHC experiments have been conducted in the Sun's upper atmosphere since it formed – and the Sun still exists.

Based upon equivalent cosmic-ray event numbers the LHC Safety Assessment Group, headed up by John Ellis of CERN's Theory Division, commented, "There is no indication that any of these previous [that is cosmic-ray produced] LHC experiments has ever had any large-scale consequences. The stars in our galaxy and others still exist, and conventional astrophysics can explain all the astrophysical black holes detected." Indeed, the argument by the Safety Assessment Group is a sound one, and while the comparison with naturally produced cosmic rays does not say that miniature black holes or strangelets won't or can't be produced by the LHC, what they do indicate is that if they are produced, then they pose no danger to our existence. Indeed, it is to be hoped that minute black holes, strangelets, and magnetic monopoles *are* produced by the LHC in order that we might begin to observe and understand such wonderful states of matter.

Looking Forward to LHCf

Compared to the other instruments situated along the LHC beam pipe, the LHCf is a relatively small detector. Rather than being measured in tens of meters it has a size that is measured in tens of centimeters. This experiment will measure collisionally produced gamma rays and pion decay products that travel very close to the beam pipe, with the specific aim of quantifying how cosmic ray collisions produce particle cascades in Earth's upper atmosphere. Two detectors, in fact, will be deployed, one located 140 m in front of the ATLAS experiment's collision point, and one 140 m behind it.

The key aim of the LHCf experiment is to compare the data derived from observed particle cascades against the various cosmic ray shower models that have been proposed over the years. The LHCf will employ two imaging shower calorimeters, and these will provide a profile of both the shower structure (Fig. 7.24) and the particle energies.

Fig. 7.24 One of the two LHCf detectors, installed along the beam pipe and ready to begin collecting data during the LHC commissioning phase. (Image courtesy of CERN)

The LHCf experiment is likely to produce some of the first new science results now that the LHC is active. Indeed, the new commissioning program suits the LHCf experiment perfectly since the two detectors cannot operate when the collider is running at full power. The LHCf research team, comprised of some 25 researchers from thirteen research groups from around the world, estimate that within just a few hours of beam time they will have gathered enough data to test the accuracy of most of the present-day cosmic ray shower models.

The King Is Dead! Long Live the King!

New discoveries about the nature of matter and the origin of the universe will be forthcoming from the LHC, and the next several years are certain to be tremendously exciting for physics and cosmology. One suggested timeline of discovery is shown in Fig. 7.25.

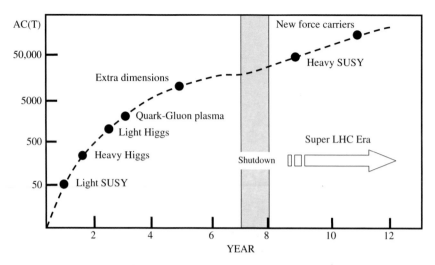

Fig. 7.25 A possible timeline for new discoveries at the LHC. Key to the discovery process is the total number of collisions accumulated, and this is the number displayed on the vertical axis in units of tera (1 million million) collisions

Given the excitement that the LHC is bound to generate over the next decade it seems almost perverse to think about its future replacement. There are, however, such projects presently being formulated at CERN and elsewhere. The next generation of advanced particle colliders is already being designed and fussed over. Beyond the LHC is the International Linear Collider (ILC), a machine once again of superlatives and once again designed with the sole purpose of pushing back the boundaries of particle physics.

The rallying cry of experimental particle physicists is always the same: "More energy – give me more energy," and it is to satisfy this craving that the ILC is being designed. Rather than being a hadron collider, however, the ILC will be an electron-positron annihilator.

As we saw in Chapter 4, protons are composed of three quarks and a veritable sea of virtual gluons and quark-antiquark pairs (Fig. 3.3). As a result of this structure there will always be some slight ambiguity about the actual quark collision energies within the LHC experiments. Although this is not an issue with respect to the fundamental discovery of new particles, it is an issue if one wishes to measure the precise properties of the new particles. One way to get around this fuzzy-collision problem is to engineer electron–positron collisions. The advantage in this case is that electrons and positrons are fundamental rather than composite particles, and their collisional energies will be more clearly defined. By way of a crude analogy, one might think of the LHC as being a shotgun and the ILC as a single-bullet rifle – the distinction being one of broad coverage over precision and localization.

The first international conference concerning the ILC was held in August 2005, and at that time its basic design and operating characteristics were defined. The essential plan is to design a collider in which the electrons and positrons are accelerated to energies of 500 GeV, and although this represents a lower total collision energy than that realized by LHC, it is more than twice that produced by CERN's LEP collider, which holds the present record of 209 GeV (LEP no longer exists, of course, since its tunnel system now houses the LHC).

The high collisional energy requirement of the ILC immediately raises a problem, since a confinement ring several hundred kilometers in circumference would be required to achieve the accelerations being sought. The construction of such a vast tunnel complex would be enormously expensive. The designers have realized, however, that a cost-effective solution to the confinement-ring problem is to use two oppositely directed linear accelerators instead, one for the electrons and one for the positrons, with the detector network being located at the central collision point.

The price tag for construct the ILC is likely to be colossal and will probably run into the many tens of billions of dollars range, but then again, just like the LHC, the ILC will require the development of numerous cutting-edge and totally new technologies. If this seems expensive, by way of comparison, the entire cost of the ILC would be but small change in the US Department of Defense's budget request for 2010 which is set at $663.8 billion – and this is dwarfed by the global military expenditure that in 2008 alone tipped $1.46 trillion.

An overview of the ILC concept design is shown in Fig. 7.26. Each of the 11.3-km long lineacs will be composed of some 8,000 superconducting cavities along which the particles will be accelerated by a traveling electromagnetic wave. Each lineac will accelerate its particle payload to energies of 250 GeV, and the collisions will be monitored by a suite of experiments located at the center point. The two damping rings at the center of the ILC (Fig. 7.26) will not be used to accelerate the particles but will be used instead to produce compact clumps of both electrons and positrons before they are induced to collide.

The ILC is currently in a design and proof-of-concept phase, and while the project has not been fully funded it might conceivably see initial construction in the early to mid-2020s. There are currently three possible locations at which the

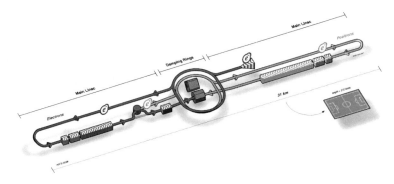

Fig. 7.26 Concept design for the ILC in which electrons and positrons will collide with energies of order 500 GeV. The entire system will be about 31 km long. (Image courtesy of CERN)

ILC might be built: CERN, Fermilab, and the Accelerator Test Facility in Japan, but no firm decision has as yet been made with respect to its eventual home.

While CERN is an active partner in the ILC program its researchers are also developing plans for another next-generation collider. The Compact Linear Collider (CLIC) has a number of similarities to the ILC. It, too, will be a linear collider, and it, too, will generate electron–positron collisions. The envisioned CLIC energies, however, will exceed those of both the ILC and LHC, with a nominal call for 3 TeV collisions. A preliminary design plan for the CLIC system is shown in Fig. 7.27. The accelerator will be some 33.6-km long and will utilize a novel two-beam system in which a drive beam is used to supply energy to the main electron and positron accelerator channels. Like the ILC, CLIC is presently in the design and concept development stage, and it is still many decades away from possible construction.

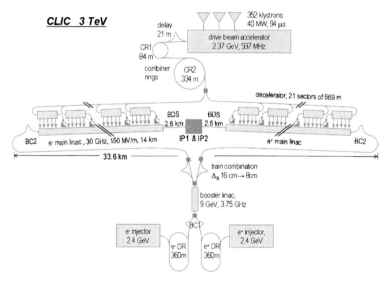

Fig. 7.27 Overview of the Compact Linear Collider (CLIC) in which it is proposed that electrons and positrons will be accelerated to collisional energies of 3 TeV. (Image courtesy of CERN)

The stage is now set, and we have indeed reached the end of the beginning. The long construction phase is over, and the LHC is ready to prove its experimental mettle. A new and ever-more detailed image of the microcosm and its reflection in the macrocosm is on the threshold of being woven. We truly live in exciting times!

> I was fervently attached to the pursuit of this subtle science and persisted in the endeavor to arrive at the truth. The eyes of opinion looked to me to distinguish myself in this beloved science. Types of machines of great importance came to my notice, offering possibilities for marvelous control.
>
> From the *Book of Knowledge of Ingenious Mechanical Devices* by Badial-Zaman Abu'l Izz Ismail ibn al-Razzaz al-Jazari (1136–1206).

Appendix A
Units and Constants

The fundamental units of length, mass, and time are taken from the System International (SI) to be the meter, the kilogram, and the second. Energies are expressed in Joules (generally) as well as in the more specialist unit of the electron volt (eV) – to be described shortly. Astronomical distances will mostly be expressed in parsecs (pc), where 1 pc approximately corresponds to the distance to Proxima Centauri, the nearest star to our Solar System (Fig. 1.1). Temperatures will be given in Kelvin, where 0 K corresponds to –273.2°C.

The two most important and fundamental physical constants appearing in this book are the speed of light and Planck's constant. In all of their experimentally measured numerical glory, we have for the speed of light $c = 2.99792458 \times 10^8$ m/s and for Planck's constant $h = 6.62607 \times 10^{-34}$ JS. The speed of light defines the greatest speed with which electromagnetic radiation (or any physical object for that matter) can travel within a pure vacuum. It is literally the absolute limiting speed within the universe. Planck's constant is fundamental to quantum mechanics and to the understanding of atomic structure. These two numbers, along with just a few others, uniquely define the structure and operation of our cosmos. Change these numbers and you literally change everything.

On occasion it will be convenient to express some quantities in terms of non-SI units. Although this practice tends to negate the whole point of having a unit system it is a commonplace practice in most scientific disciplines. Among the key unit conversions are: 1 eV $= 1.602176 \times 10^{-19}$ J and 1 parsec (pc) $= 3.085678 \times 10^{16}$ m $= 206,265$ AU $= 3.261631$ light years (ly). The first unit conversion, as stated earlier, relates specifically to atomic physics and is defined as being the energy acquired by a single, unbound electron after being accelerated through an electrostatic potential difference of 1 V. The exact physical details need not concern us here; all we need to know is that there is a specific definition for the eV energy unit. The second conversion relates the parsec to the meter and light year. The parsec is defined (see Fig. 6.2) as being the distance at which the half angle of the apparent positional shift on the sky, as seen from Earth's orbit over a 6-month time interval, is equal to 1 s of arc (an angle equivalent to 1/3,600th of a degree).

Although the parsec is a fundamental measurement of stellar distance, based upon a measured time interval (6 months) and a measure of Earth's orbital

M. Beech, *The Large Hadron Collider*, DOI 10.1007/978-1-4419-5668-2,
© Springer Science+Business Media, LLC 2010

semi-major axis – a distance defined as being 1 astronomical unit (AU) – the light year is perhaps more familiar, although it is by no means any more meaningful in conveying the measure of astronomical distance to the human mind. The light year, of course, is defined as the distance traveled by light, traveling at speed c, in 1 year. Since there are 365.2425 days in a (Gregorian) year, with each day being by definition 86,400 s long, so 1 ly $= 9.460536 \times 10^{15}$ m.

In addition to the basic units described above it will be convenient to express numbers in terms of powers of ten: 10^P, where P is some number. In this manner, for example, the number one million (1,000,000) is written as 10^6, and such a power is described by the word mega and represented by the symbol M.

The SI system recognizes a variety of numerical shorthand names and symbols and these are given in Table A.1 below. Accordingly, for example, distances on the cosmological scale will typically be cast in terms of mega (one million) and giga (1,000 million) parsecs, while characteristic atomic sizes, such as say the diameter of the first electron orbit of the hydrogen atom, will be written in terms of nano (1,000 millionths) meters. The characteristic size of an atomic nucleus, as we shall see in Chapter 1, is expressed in terms of femto meters.

Table A.1 Powers of ten and their shorthand names and symbols

Power of ten	Number	Name and symbol
10^{-15}	0.000000000000001	Femto (f)
10^{-12}	0.000000000001	Pico (p)
10^{-9}	0.000000001	Nano (n)
10^{-6}	0.000001	Micro (μ)
10^{-3}	0.001	Milli (m)
10^{0}	1.0	Unity
10^{3}	1,000	Kilo (K)
10^{6}	1,000,000	Mega (M)
10^{9}	1,000,000,000	Giga (G)
10^{12}	1,000,000,000,000	Tera (T)
10^{15}	1,000,000,000,000,000	Peta (P)

Appendix B
Acronym List

The author happily admits to a great dislike of acronyms, and especially the highly contrived ones beloved of particle physicists. Bowing, albeit reluctantly, however, to convention and the greater wit of others, acronyms have been used throughout the text, and a summarizing list of them is given below.

ADMX	Axion Dark Matter Experiment
AGB	Asymptotic Giant Branch
ALICE	A Large Ion Collider Experiment
ATLAS	A Toroidal LHC ApparatuS
AU	Astronomical Unit
BOOMERanG	Balloon Observations Of Millimetric Extragalactic Radiation and Geophysics
CERN	Conseil Européen pour la Recherche Nucléaire
CDF	Collider Detector at Fermilab
CDM	Cold dark matter
CDMS	Cryogenic Dark Matter Search
CHAMP	CHArged Massive Particle
CLIC	Compact LInear Collider
CMB	Cosmic microwave background
CMS	Compact Muon Solenoid
COBE	Cosmic Microwave Background Explorer
COSMOS	Cosmic Evolution Survey
COUPP	ChicagO Underground Particle Physics laboratory
CP	Charge-parity
DARMA	DARk MAtter experiments
ECAL	Electromagnetic CALorimeter (part of CMS)
ESA	European Space Agency
ESO	European Southern Observatory

GLAST	Gamma-ray Large-Area Space Telescope. Spacecraft renamed Fermi Gamma-ray Space Telescope after successful launch in June of 2008.
GUT	Grand Unified Theory
HCAL	Hadron CALorimeter
HR	Hertzsprung–Russell (diagram)
ICG	Intercluster gas
ILC	International Linear Collider
ISM	Interstellar medium
ISW	Integrated Sachs–Wolfe effect
ITS	Inner Tracking System
JDEM	Joint Dark Energy Mission
LBV	Luminous Blue Variable
LDEF	Long Duration Exposure Facility
LEP	Large Electron–Positron collider
LHC	Large Hadron Collider
LHCb	Large Hadron Collider beauty experiment
LHCf	Large Hadron Collider forward experiment
LIBRA	Large sodium Iodide Bulk for RAre processes experiment
LINEAC	LINEar ACcelerator
LSP	Lightest Supersymmetric Particle
MACHO	Massive Compact Halo Object
M/L	Mass to light ratio
MoEDAL	Monopole and Exotics Detector At the LHC
MOND	Modified Newtonian Dynamics
MSSM	Minimum Supersymmetry Standard Model
NASA	National Aeronautics and Space Administration
NGC	New General Catalog
OPERA	Oscillation Project with Emulsion-tRacking Apparatus
PAMELA	Payload for Matter-Antimatter Exploration and Light-Nuclei Astrophysics
PEP	Pauli Eexclusion Principle
PMT	PhotoMultiplier Tube
PP	(The) proton–proton (chain)
PS	Proton Synchrotron
PSB	Proton Synchrotron Booster
QCD	Quantum ChromoDynamics
QED	Quantum ElectroDynamics
QGP	Quark–Gluon Plasma
RF	Radio frequency
RHIC	Relativistic Heavy Ion Collider
RICH	Ring Imaging Cerenkov Detector
SIMP	Strongly Interacting Massive Particle
SLAC	Stanford Linear ACcelerator
SMH	Strange Matter Hypothesis

SNO	Sudbury Neutrino Detector
SPS	Super Proton Synchrotron
SUSY:	Supersymmetry
2dFGRS	The 2-degree Field Galaxy Redshift Survey
TOE	Theory Of Everything
TOF	Time Of Flight
TOTEM	ToTal Elastic and diffraction cross-section Measurement
TPC	Time Projection Chamber
TRD	Transition Radiation Detector
WIMP	Weakly Interacting Massive Particle

Appendix C
Glossary of Technical Terms

Baryon	Normal matter. The class of hadrons made of three quarks.
Black hole	A gravitationally collapsed object.
Blackbody	An idealized object that absorbs all incident electromagnetic radiation but also radiates a very specific temperature and wavelength-dependent thermal radiation spectrum.
Blueshift	Apparent reduction in the wavelength of light due to motion towards the observer (see the Doppler effect).
Boson	The general name for particles having integer spin.
Cerenkov radiation	Electromagnetic radiation emitted when a charged particle traverses a medium with a speed faster than the speed of light in that medium.
Cosmic ray	A charged particle (e.g., proton or electron) moving through interstellar space at close to the speed of light.
CP symmetry	This rule states that the laws of physics should remain the same if a particle is swapped with its antiparticle (so-called C symmetry) and if left and right handedness are interchanged (so-called P symmetry).
CP Violation	The violation of CP symmetry that resulted in a matter-dominated universe.
Dark energy	Hypothetical energy of unknown origin that causes the universe to expand at an accelerating rate.
Dark matter	Matter of unknown origin and structure that permeates the entire universe but does not interact with electromagnetic radiation. Its existence is revealed through its gravitational effect on baryonic matter.

Degenerate gas	A gas in which the Pauli Exclusion Principle operates.
Doppler effect	An apparent change in the measured wavelength of light due to a relative motion between the observer and the emission source.
Electromagnetic radiation	A self propagating wave that travels at the speed of light and is composed of oscillating magnetic and electrical fields.
Electromagnetic force	The fundamental force that acts between electrically charged particles.
Electron	A lightweight, fundamental particle that carries a negative charge.
Event horizon	Boundary around a black hole at which the escape velocity is equal to the speed of light.
Fermion	The general name for particles having half-integer spin (e.g., the electron).
Gamma ray	A very high energy form of electromagnetic radiation (a very high frequency, short wavelength photon).
Gluon	A massless particle that carries the strong nuclear force.
Graviton	A hypothetical, massless particle that carries the gravitational force within the framework of quantum gravity.
GUTs	Grand Unified Theories. Theories that predict the eventual coalescence of the electromagnetic, weak nuclear, and strong nuclear forces at very high energies.
Hadrons	Particles composed of quarks and antiquarks.
Hertzsprung–Russell diagram	The diagram in which stellar surface temperature is plotted against stellar luminosity. The diagram was first constructed by Eijnar Hertzsprung and Henry Norris Russell.
Leptons	Particles that do not feel the strong force and have half integer spin.
Luminosity	In physics this is the number of particles per unit area per second. In astronomy it is the total amount of electromagnetic energy that a star radiates into space per second.
Magnetic monopole	Particle supporting a single magnetic polarity.
Main sequence	The locus in the Hertzsprung–Russell diagram of those stars generating energy by hydrogen fusion reactions within their interiors.
Meson	The class of hadrons composed of a quark-antiquark pair.

Muon	A massive version of the electron.
Neutrino	An electrically neutral particle that interacts weakly with matter and is a member of the lepton family.
Neutron	Electrically neutral partner to the proton in an atomic nucleus. Composed of two down and one up quark.
Neutron star	A stellar mass object of very small size supported against gravitational collapse by degenerate neutrons.
Nucleus	The central component of an atom containing protons and neutrons.
Pauli Exclusion Principle	A quantum mechanical condition that forbids any two fermions from having the same quantum state at the same location and time.
Photon	A massless particle that carries the electromagnetic force.
Pion	A meson composed of an up or a down quark along with its antiquark partner.
Pulsar	A rapidly rotating, highly magnetic neutron star that generates a pulse of electromagnetic radiation each time its magnetic pole is brought into the observer's line of sight.
Proton	The positively charged component of an atomic nucleus. Composed of two up and one down quark.
Quark	Fractionally charged "seed particles" that makeup hadrons.
Red giant	A star that has exhausted hydrogen within its central core, and is undergoing structural readjustment prior to the initiation of helium fusion reactions. Such stars are characterized by having very large radii and low surface temperatures.
Redshift	An apparent increase in the wavelength of light due to motion away the observer (see the Doppler effect).
Rest energy	The energy E associated with a quantity of matter M. Described by Einstein's formula $E = M c^2$, where c is the speed of light.
Space velocity	The speed with which a star or stellar system is moving through space.
Spin	Measure of the intrinsic angular momentum of a particle.
Standard candle	An astronomical object (star, galaxy, etc.) of known luminosity.

Strong force	The fundamental force that binds quarks and anti-quarks together to make hadrons. The strong force is described by QCD theory.
SUSY	Supersymmetry. A theory uniting fermions and bosons, where every known particle has a massive companion particle whose spin differs from it by one half.
Synchrotron radiation	Radiation produced by a charged particle moving along a curved path – such as in a circular synchrotron accelerator.
Uncertainty Principle	A quantum mechanical condition that sets limits upon the accuracy with which the position and momentum of a particle can be simultaneously measured.
Virial theorem	A theorem that relates the time-averaged value of the kinetic energy of a stable many-bodied system to that of the total potential energy.
White Dwarf	A stellar mass object of small size supported against gravitational collapse by the outward pressure due to degenerate electrons.

Index

A

ADMX, 157–158, 225
ALICE, 44, 48, 62, 67, 79, 112–115, 151, 225
Alpha particle, 14–15
Alpha, Ralph, 100
Anderson, Carl, 73
Andromeda galaxy, 2–4, 130–133,
    137–138, 184
Apelles of Cos, 40
Aquinas, Thomas, 115, 119
Archimedes, 39–40
Atom, 1, 5–9, 11–22, 24–26, 29–37, 40, 49,
    53, 62, 73, 75, 77, 87, 91, 96–97, 101,
    112, 123, 148, 152–154, 173, 180, 185,
    202, 205, 207, 210, 215–216, 224, 231
A Toroidal LHC ApparatuS (ATLAS), 42, 48,
    62, 64–67, 83, 112, 210–211, 219, 225
Axion, 116–117, 149, 157, 225
Aymar, Robert, 45, 47

B

Bacon, Francis, 189
Balmer, Johann Jakob, 19–20
Balmer lines, 20
Baryonic mass, 105
Becquerel, Antoine Henri, 14
Berners-Lee, Tim, 54
Beta decay, 25–26, 76–77, 81–82, 103, 215
Beta particle, 14, 25
Blackbody radiator, 16–17, 21–22, 34, 100
Black Hole, 33, 37, 49–50, 149, 163, 174, 185,
    193, 195–197, 203–211, 213–214, 218,
    229–230
BlackMax, 210
Bohr atom, 15–21
Bohr, Niels, 19–20, 22, 86
Bose, Satyendra, 87
Brahe, Tycho, 177
Brout, Robert, 78

Bubble chamber, 63, 152–153
Bullet Cluster, 146–147, 160

C

Carroll, Lewis, 82, 225
Cavendish Laboratory, 13, 49, 73
ΛCDM model, 182, 184
Cepheid variable, 130
Cerenkov radiation, 118, 214, 229
CERN, 28, 40–56, 60–61, 63–70, 73, 78, 80,
    83–85, 88–91, 112–115, 117–118,
    162–163, 189–190, 207, 210–212,
    219–222, 225
Chadwick, James, 15, 73
CHAMP, 153–154, 225
Chandrasekhar limit, 174, 176, 178
Chandrasekhar, Subrahmanyan, 174–176, 178
Chandra X-ray telescope, 146, 201, 204
Chicagoland Observatory for Under-
    ground Particle Physics (COUPP),
    151–154, 225
Chomolungma, 2, 54
Coles, Peter, 106
Coma cluster, 134–136
Compact Linear Collider (CLIC), 222, 225
Compact Muon Solenoid (CMS), 43, 46,
    62, 64, 66–69, 79, 83–84, 87–88,
    112, 225
Copernican principle, 93, 186
Copernicus, Nicolaus, 31, 91–92, 186
Cosmic microwave background (CMB),
    99–103, 107–109, 181, 225
COsmic Microwave Background Explorer
    (COBE), 101–102, 225
Cosmic ray, 26, 50–51, 73, 149, 151–152,
    155, 157, 163, 202–203, 207–208, 213,
    216–219, 229
    shower, 216, 219
Cosmological constant, 164, 182

Cosmological principle, 93–94, 97, 186
Cosmological redshift, 180
COSMOS, 5–6, 9, 40, 71, 91, 95–96, 116, 132,
    145, 146, 148, 162, 164, 223, 225
Cowan, Clyde, 73
CP violation, 116, 149, 229
Critical density, 96–99, 103, 105,
    109–110, 168
Cronin, James, 116
Curtis, Heber, 129, 167

D

Dark energy, 73, 99, 103, 109, 128, 162–187,
    190, 226, 229
Dark fluid, 183
Dark matter (DAMA), 73, 84, 90, 98, 103,
    117, 121–164, 181, 183, 185, 190, 217,
    225, 229
    halo, 127, 141–142, 161
Dark stars, 160–162
Davis, Raymond, 26–27
De Broglie, Louis, 29, 56
2dFGRS, 133–134, 227
Digges, Thomas, 92–93
Dimopoulos, Savas, 87
Dirac, Paul, A. M., 87, 211–213
Dodgson, Charles, 151–152
Doppler Effect, 137, 229–231
Dreyer, John, 129
Dürr, Stephen, 76

E

Eddington, Arthur, 38–39, 190
Einstein, Albert, 17, 21–22, 57–58, 75–76, 87,
    95–96, 107, 142–143, 159, 164, 180,
    183, 191, 204, 210, 231
Eliot, Thomas Stearns, 90
Ellis, John, 84, 218
Empedocles, 6
Englert, François, 78
Euclid of Alexandria, 97, 157–158
Exclusion Principle, 24, 34, 87, 173, 195, 226,
    230–231

F

Fermi, Enrico, 25, 87
Fermilab, 49, 72–74, 84–86, 152, 154,
    161–162, 212, 222, 225
Feynman diagram, 79–84
Feynman, Richard, 24, 80
Fine structure constant, 25
Flatness problem, 106–107
Ford, Kent, 139
Friedman, Alexander, 96–98, 101, 107

G

Galaxies
    barred spiral, 131
    elliptical, 10, 131–132, 168
    irregular, 128, 131–132
    lenticular, 131, 136
    local group, 132–133, 167, 169, 180,
        184–185, 205
    magellanic clouds, 132, 177, 201
    sombrero, 137–138
    spiral, 2, 18, 63, 129–132, 135, 139–142,
        168, 196, 216
    whirlpool, 129
Galilee, Galileo, 98
Gamow, George, 100
Geiger, Hans, 14–15
Gell-Mann, Murray, 29–30
Gilbert, Walter, 78
Gillmor, Helen, 50
Glashow, Sheldon, 89
Globular cluster, 127, 166–167
Grand Unified Theories (GUTS), 88–89, 111,
    212, 226, 230
Gravitational lensing, 142–147, 149, 160–161
Graviton, 75, 210, 230
Guth, Alan, 106, 110

H

Hawking radiation, 185, 208–210
Hawking, Stephen, 82, 85, 185, 208–210
Heisenberg, Werner, 22–23, 25, 76, 108, 208
Herschel, William, 2, 121–122, 124, 126,
    128–129, 136
Hertzsprung-Russell diagram, 171, 191, 230
Higgs field, 58, 76, 78–79, 82
Higgs particle (boson), 57–58, 64, 70, 79
Higgs, Peter, 70, 78, 84
Holbein, Hans, the younger, 144–145
Homestake Mine, 27
Horizon problem, 106–108
Howard, Georgi, 87
Hoyle, Fred, 95–96
Hubble's constant, 93, 97–98, 109, 167–168
Hubble, Edwin Powell, 92, 130–131, 164, 167
Hubble Deep Field, 38–40
Hubble's law, 93–94, 107, 168, 184, 186
Hubble Space Telescope, 38, 129, 138,
    145–146, 168
Hubble time, 135
Hubble's tuning fork, 131
Humason, Milton, 167
Hypernova, 197, 216

I

Inflation, 105–110, 181
Innis, Robert, 166
International Linear Collider (ILC),
220–222, 226
Internet, 54
Interstellar dust, 123–125, 130, 138
Interstellar medium (ISM), 123–125, 130, 178,
185, 192, 198, 214, 216, 226

J

Joyce, James, 29

K

Kant, Emmanuel, 85, 130
Kapteyn, Jacobus, 126, 136
Kelvin, Lord, 13
Kepler, Johannes, 9–12, 140, 177
Khvolson, Orest, 142

L

Large Electron-Positron (LEP), 41, 44, 57, 60,
73, 85–86, 217, 221, 226
Lederman, Leon, 73, 79
Lemaître, Georges, 96
LHCb, 44, 62, 67, 116–119, 213, 226
LHCf, 62, 218–219, 226
LINEAC, 52, 226
Lovelock, James, 5

M

Magnetic
field, 21, 59–60, 63–64, 66, 74, 79,
113, 117, 149, 154, 157, 193, 195,
212–213, 216
monopoles, 49–50, 211–213, 218, 230
steering, 47, 60–61
superconducting, 44, 55, 60–62, 64, 66
Main sequence lifetime, 33–34, 190–191
Marsden, Ernest, 14
Massey, Richard, 145
Massive Compact Halo Objects (MACHO),
149, 162, 226
Maxwell, James Clerk, 17–18
Mcluhan, Marshall, 125
Meteorite, Gibeon, 215
Milgrom, Mordechai, 159
Milky Way galaxy, 2, 39, 124–127, 132–133,
136–137, 140, 149, 166, 177, 180,
203–204
MOdified Newtonian Dynamics (MOND),
159–160, 226

MSSM, 87, 226
Myers, Steve, 48

N

Nagaoka, Hantaro, 15, 18
Neutralino, 87–88, 147–149, 155
Neutrino
discovery, 73
flavors, 26–28, 148
OPERA, 28, 226
solar, 26–28, 148
Neutron star, 37, 149, 163, 185, 192–201, 208,
214, 231
Newton, Isaac, 56, 159–160, 194

O

Objects
Abell 39, 128
Abell 2199, 105
Abell 2218, 143–144
Asteroid (12343), 205
Barnard-68, 123
Betelgeuse, 34–35
3C58, 198–201
α-Centauri, 160, 165–166
Crab nebula, 200
61 Cygni, 166
Cygnus X-1, 196
1E 0657-56, 146
Eta Carina, 197
G1.9+0.3, 178
Geminga, 193–194, 196
IK Pegasus, 177–178
LBV 1806-20, 169
M57, 137, 173
NGC2808, 166
NGC 3115, 136
NGC 3594, 137
NGC 4526, 169
NGC 4565, 139
Proxima Centauri, 3, 132, 160, 166, 223
RXJ1856.5-3754, 197–203
Sgr A*, 204–205, 210
Sirius A, 173–174
Sirius B, 173–174
SN 1572, 177
SN1987A, 201
SN 1994d, 169
Vega, 165–166
Vela Pulsar, 194
WD 0137-349, 138
Oort cloud, 123, 160
Oort, Jan, 126, 136
Orwell, George, 86

P

PAMELA, 155–157, 217, 226
Parallax, 126, 165–167
Pauli, Wolfgang, 23–24
Penzias, Arno, 99–101
Periodic table, 8, 31–32, 36, 192
Perl, Martin, 73
Perlmutter, Saul, 180
Planck's constant, 4, 16, 22, 86, 223
Planck mass, 206–208, 210
Planck, Max, 4, 15, 17, 22
Planck scale, 89, 205–206
Planetary nebula, 37, 128, 172–173, 191
Plato, 5–8, 10–11, 72
Platonic solids, 6–7, 10–11
Proton-proton chain, 33, 77, 171

Q

Quantum chromodynamics (QCD), 76, 113, 116, 226
Quantum fluctuations, 108–109, 208–209
Quantum gravity, 4, 75, 89, 111, 206, 230
Quark-Gluon Plasma (QGP), 111–113, 115, 199, 220, 226
Quark star, 199, 201, 214

R

Recombination, 100–101, 105, 109, 173
Regina, 2
Reines, Frederick, 73
RHIC, 113, 154, 201–202, 217, 226
Rosse, Third Earl of, 129–130
Rubbia, Carlo, 73, 89
Rubin, Vera, 139–141
Ruskin, John, 2
Rutherford, Ernest, 13–15, 18, 21, 29, 49, 73, 86
Rutherford scattering experiment, 14–15

S

Salam, Abdus, 89
Schmidt, Brian, 180
Schrödinger, Erwin, 23, 25
Schwarzschild, Karl, 204
Schwarzschild radius, 204–205, 207, 209–210
Shapley, Harlow, 127, 130
Slipher, Vesto, 137–138
Smith, Sinclair, 136
Socrates, 5
Sodium ethyl xanthate (SEX), 32
Stanford Linear Accelerator Center (SLAC), 73, 226
Standard Model, 11, 30–31, 40, 70–90, 112, 116, 147, 182, 190, 206, 213, 226
Stoney, Johnstone, 13

Strangelets, 49–50, 201–203, 213, 218
Strange matter, 199, 202, 226
    hypothesis, 202, 226
Strong nuclear force, 74–75, 111, 119, 211, 230
Struve, Otto, 136
Sudbury Neutrino Observatory, 27
Super Proton Synchrotron (SPS), 43–44, 53, 56, 62, 73, 96, 113, 212, 227
Supernova
    remnants, 178, 193–194, 200
    type I, 168, 170, 175–180, 182–183, 186
    type II, 170, 177, 200
Supersymmetry (SUSY), 84, 86–88, 147, 220, 227, 232

T

Tevatron, 74, 85, 161
Thomson, Joseph John, 13, 73, 101
Thomson scattering, 101
Thomson, William, 13
TOTal Elastic Measurement (TOTEM), 62, 66–67, 70, 227
Triple-alpha reaction, 34–35

U

Ultraviolet catastrophe, 16
Uncertainty principle, 23, 25, 76, 108, 232

V

Van de Hulst, Hendrik, 138
Van der Meer, Simon, 73, 89
Virgo cluster, 133, 136, 169
Virial theorem, 135, 232
Vitch, Val Logsdon, 116
Von Seeliger, Hugo, 126

W

Weak nuclear force, 57, 74, 76–77, 88–89, 111, 148
Weinberg, Stephen, 89
White dwarf, 34, 37, 138, 149, 170, 172–178, 185, 191–192, 232
Wilson, Robert, 99
WIMP, 148–152, 161–162, 227
Wirtz, Carl, 167
Witten, Edward, 202

Y

Y(4140), 160–162

Z

Zwicky, Fritz, 135